RIVERINE DREAMS

ALSO BY GEORGE FRAZIER

The Last Wild Places of Kansas: Journeys into Hidden Landscapes

RIVERINE DREAMS

Away to the Glorious and Forgotten Grassland Rivers of America

GEORGE FRAZIER

THE UNIVERSITY OF CHICAGO PRESS

Chicago and London

The University of Chicago Press, Chicago 60637
The University of Chicago Press, Ltd., London
© 2025 by The University of Chicago
Published 2025
Printed in the United States of America

34 33 32 31 30 29 28 27 26 25   1 2 3 4 5

ISBN-13: 978-0-226-83879-3 (cloth)
ISBN-13: 978-0-226-83880-9 (ebook)
DOI: https://doi.org/10.7208/chicago/9780226838809.001.0001

Library of Congress Cataloging-in-Publication Data

Names: Frazier, George, 1966– author.
Title: Riverine dreams : away to the glorious and forgotten
    grassland rivers of America / George Frazier.
Description: Chicago : The University of Chicago Press, 2025. |
    Includes bibliographical references and index.
Identifiers: LCCN 2025003883 | ISBN 9780226838793 (cloth) |
    ISBN 9780226838809 (ebook)
Subjects: LCSH: Rivers—Great Plains. | Stream conservation—
    Great Plains. | Stream conservation—United States. |
    Grasslands—Great Plains. | Grassland restoration—Great
    Plains. | Grassland restoration—United States.
Classification: LCC QH541.5.S7 F73 2025 | DDC 577.6/40978—
    dc23/eng/20250403
LC record available at https://lccn.loc.gov/2025003883

♾ This paper meets the requirements of ANSI/NISO Z39.48-1992
(Permanence of Paper).

For my parents, Sara and Bill Frazier

Take me away to nowhere plains.

BLACK FRANCIS

CONTENTS

AUTHOR'S NOTE

This is a work of nonfiction, reported as events happened and without embellishment. The trips are not always presented in strict chronological order and in a few cases names have been changed to protect identities. Any mistakes, omissions, or inaccuracies are the fault of the author.

The author and publisher wish to gratefully acknowledge that part of chapter 3 was derived from my essay "Wild Winds on the Grand," originally published in *The New Territory: The Magazine of the Lower Midwest* 8 (2019): 45–53.

Map design by Christina Frazier.

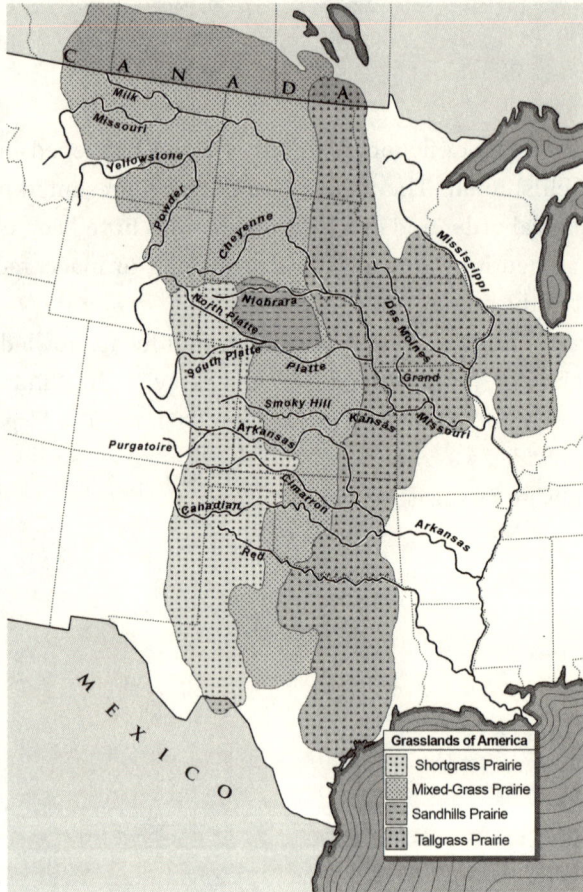

MAP 1 The Grasslands and Grassland Rivers of America

CHAPTER 1

RIVERINE DREAMS

I believe America is at the beginning of a grassland river revival. A revival in Montana, where a motley alliance is stitching together a prairie reserve of such continental scale that clouds of buffalo, elk, and antelope might once again thunder across the breaks of the most storied grassland river on earth. A revival in Colorado, where meltwater torrents sometimes strike terror in the hearts of newly arrived techies, craft beer fanatics, and legalized marijuana entrepreneurs who are remaking Front Range metropoles into their own reimagined visions of the West. A revival in Wyoming, the Dakotas, and Nebraska, a state so stricken with heartbreaking rivers that I wouldn't blame anybody who flings this book aside to set out for its crystalline waterfalls, sandhill crane mobs, and mysterious upwelling springs that tap deep into the heart of the Ogallala. A revival in Iowa, Oklahoma, Kansas, and Missouri, that blockbuster of river states, where a festival of paddlers provisioned with pharmaceutical-grade caffeine, vegan pemmican, and living wills embarks each summer on the world's longest nonstop ultra-marathon river race. And a revival along thousands of miles of deceptively ordinary rivers across the prairies and plains, where fifty million people go about their lives in the ruins of North America's once vast interior grassland.

The grasslands of the trans-Mississippi West have gone by many names. Prairie (from the Old French *praerie* or meadow and not, as I'd once thought, a prayer to the spirit who stalks the horizon), steppe, pasturage, Great Plains, High Plains, plain plains, wolds, downs, balds, barrens, flatlands, seas of grass, llanos, unbroken sod, sagebrush, shortgrass, mixed grass, sweetgrass, the land of tallgrass and cheap beer, rugged pastures sparkling and empty as Buddha-mind, ancient lawns that to many wilderness seekers sound about as promising as karaoke night at a mime convention. Appreciating the beauty of this spare, self-deprecating landscape is like meditation or long-distance running: proficiency takes time and accrues imperceptibly. Then suddenly you're all in.

Grasslands didn't always blanket middle America. Plate tectonics keep shuffling the continents around, and worldwide climate has ping-ponged between the extremes of "snowball earth"—when only a thin equatorial ribbon remained ice-free—and eons when crocodiles swam at the North Pole. Across deep time, every square inch of *terra familiaris* has worn a coat of many ecosystems. Since the ice sheets of the last glacial period retreated to hibernate in the far north about twelve thousand years ago, much of the land between the Rockies and Appalachia has settled into a climate where rainfall is insufficient to support perennial forests but just enough to avert outright desertification. A Goldilocks zone for grasslands.

The precise boundaries of these prairie provinces—which can be further subdivided into shortgrass, mixed-grass, and tallgrass regions—have ebbed and flowed like an inland tide, as wavelike throngs of hooved specialists evolved to convert grass into quivering biomass. Fire, whether sparked by lightning or set intentionally by Native Americans who hewed their environs for game and the growth of edible plants, further cleansed the lands east of the Rockies and west of the eastern hardwoods—a mostly treeless, but peopled, wilderness up until the arrival of European explorers in the sixteenth and seventeenth centuries.

It took four generations to destroy.

The Indian Removal Act of 1830 and subsequent treaties forced Native peoples from the east onto grassland reservations. US Army campaigns to subjugate Native peoples of the West in the late nine-

teenth century, and their subsequent removal to the Oklahoma Indian Territory, snuffed out cultural fires across most of the grasslands. In the tallgrass prairie states especially, early land surveys
indicate woody species had already begun to overtake prairies
before homesteaders arrived. The eradication of buffalo further
stressed grasslands that had coevolved with the herds.

"Sod-breakers" finished the job. In eighty years, homesteaders
and farmers plowed more than 96 percent of the original prairies.
Cultural revolution sowed environmental revolution. The people
of the prairies and plains spoke at least forty-five distinct lan-
guages in the late 1700s, twenty in the late 1800s, and mostly
one—English—in the late 1900s. The magnificent biodiversity of
the post-Pleistocene bison kingdom vanished. Gone were prairie
dog colonies larger than European nation states. Gone were the
fathomless herds of buffalo, elk, and antelope. Gone were plains
wolves and grizzly bears that had followed those herds for millennia. From a wild lands perspective, by the time I was growing up
in the once-prairied suburbs outside of Kansas City, the quintessential American dreamscape had been farmed over, diminished
to nostalgia, a "flyover" country where the deer and the antelope
played.

But America is still a nation of dreamers. After centuries taming the landscape, attitudes about wild lands began to shift in the
twentieth century. Federal environmental policy followed suit with
the Wilderness Act of 1964, the Environmental Protection Agency,
the Endangered Species Act, and the Clean Water Act. People donated time and money to the Sierra Club, Nature Conservancy, and
Wilderness Society, organizations that purchased land and protected threatened ecosystems. Millions of wilderness acres were enrolled in what seemed like an ironclad federal protectorate. Rather
than converting nature into a garden, some Americans began to
value nature *as* garden, and fought to save wilderness. The American bison, California condor, northern gray wolf, bald eagle, black-
footed ferret, and American alligator clawed their way back from
the edge of extinction. But protecting endangered ecosystems—

entire communities of interconnected life that coevolved over time spans too vast for human comprehension—is a recent concept. The Florida Everglades, the Atchafalaya swamp in Louisiana, California redwood forests, the last remaining mountain enclaves of giant sequoia, and the Boundary Waters wilderness in northern Minnesota were small pseudo-victories, ecosystems that once teetered on the brink, and to varying degrees managed to stave off modernity through complicated salvage and rewilding operations.

But the prairies could become the passenger pigeons of American ecosystems. Less than 5 percent of the original North American grasslands remain, and prairie restoration is a young science with a kind of Turing test. In 1950, British cryptanalyst and mathematician Alan Turing, who helped crack the German Enigma code during World War II, proposed an experiment to decide whether a machine could think like a human. Basically, if several people spend a few hours vibing with an artificial intelligence and all of them believe it's a person, then the artificial intelligence passes the test. In the botanical version of the Turing test, if plant scientists spend all morning studying an artificial prairie and all of them believe it's unplowed, then the artificial prairie passes the test. As of 2025, no artificial intelligence or artificial prairie has done so.

Never-plowed prairies are complex biological communities, mélanges of forbs and grasses that, inch per inch, exceed Brazilian rainforests in biodiversity. One Missouri prairie had forty-six different kinds of plants growing in a twenty-by-twenty-inch transect, the most botanically diverse quarter meter ever documented. A pasture enrolled in the USDA's Conservation Reserve Program might support fewer than a dozen species. The holy grail in prairie restoration is to sow eighty or more seed varieties and get eighty or more thriving species a decade later. Until recently, such sowings usually resulted in fewer than twenty. However, botanists, soil experts, and other scientists are working to solve this basic prerequisite of grassland rewilding. Once they do, it could be possible to re-create prairies and stitch together larger grasslands, eventually at scales sufficient to handle migrating bison, elk, antelope, and someday the predators that stalked them. Population declines in the Great Plains that once spurred talk of a "buffalo commons" have accelerated. A heartland with agriculture, ranching, cities, *and* grassland wilderness might resonate with future middle Americans.

One day in college, I pulled off the road to find out why a crowd had gathered across the highway from a well-known natural landmark on the outskirts of Lawrence, home of the University of Kansas. Elkins Prairie was a rare deep-soil prairie that harbored two federally threatened plants—Mead's milkweed and western prairie fringed orchid. Travelers on the California Trail probably stopped there for water from a small seep spring that still flows. Botanists had cataloged a litany of fragile species that only grow in the highest quality undisturbed prairies, their vast root networks unmolested since the time of the glaciers.

But Elkins Prairie had become embroiled in politics. The state planned to build a new highway near the site. After fumbled attempts by environmental groups and the county to purchase the prairie (the Nature Conservancy had offered 50 percent of the going rate for nearby properties), the landowner decided to plow the ancient grassland to make it "more productive." The crowd swelled as news got out about the intentional destruction of one of the most beloved natural areas in Kansas.

Standing there watching the tractor convert ten thousand years of natural history into a memory lit a fire in my heart for the grasslands of the Kansas River valley. I'd never equated them with the wilderness of the American West that I'd grown to love by scrambling up Colorado fourteeners, backpacking in the Wyoming Tetons and Never Summer Mountains, bike-camping down the coast of Washington and Oregon, and hanging out in Northern California during "Redwood Summer," when environmentalists chained themselves to trees to protest the logging of old-growth coastal redwoods. Whether they mourned or believed the landowner had been slighted and was within his legal rights to do whatever he wanted with his land, everyone standing along the highway on that gray November day felt passionate about what was happening. Elkins Prairie was a wild place that *mattered*, in the way that Western wild places mattered. When a small journal published my article about the tragedy, I began freelancing pieces about prairies and the natural landscapes of the Great Plains and the Midwest, and I eventually wrote a book about a hidden map of wild places in remote Kansas locales. If, as geographer Yi-Fu Tuan said, space plus culture equals place, grasslands became my place.

Now, years later, gravity seemed to be pulling my family away from the Midwest. Chloe, our daughter, was about to finish high school and leave for college; Christina, my wife, had transitioned to a career with more promising opportunities in Florida and along the East Coast; and my job as a software architect had me traveling the long sides of a triangle that joined San Jose, Boston, and Bangalore—far-flung hubs of the tech world, none with midland prairies.

The timing of it was frustrating. When it came to knowing prairies in the deep way I wanted to, I'd always thought I was born too late, that the part of me that yearned for midland wilderness was destined to starve, sickened on a diet of cornfields, wheat farms, cul-de-sacs, and Walmart. But with science dangling new threads of hope for repairing lost grasslands, maybe leaving the Midwest would uproot us from prairie soils just at the dawn of a new era.

At the midpoint of a long hike that winter, I spread out my tarp to rest in a cemetery at the top of a steep bluff above Golden Road, the closest thing Kansas City has to a proper river road along the Kansas, or Kaw River, as it's often called. I'd been dribbling around the idea of a grand grasslands barnstorm, one last meaningful foray across the prairies. Not to troll for relicts—others had mapped lost canyons of high plains soapweed and prairie dogs, large tracts of tallgrass cattle pastures in the Kansas Flint Hills; the last vestiges of unplowed wilderness clinging to life in old graveyards, along rail embankments, and at the tops of glaciated hills—or to marinate in Western nostalgia: Lewis and Clark, Buffalo Bill Cody, cowboys and Indians. I was looking for something less obvious that I didn't understand. A chance to experience the grasslands before their fate was known, whether scientists would instigate a prairie resurgence, or, perhaps more likely, indifference, invasive species, and climate change would lap away at the last tiny islands of prairie until they disappeared forever.

After hiking all morning, it felt like somebody had poured concrete into my hip flexors. But lying back in the winter sun uncorked miles of pent-up endorphins and a warm radiance poured over me. Above the little cemetery, a solitary bald eagle—once a rare, endangered symbol, a member of Rachel Carson's brigade, now a commoner—soared in a great circle. Green Berry Chance, born in Cherokee, North Carolina, in 1813, chose the bluff for his family cemetery. He might have known about the stories of Native Americans buried there. But looking west toward the cold horizon, it was obvious to me that he chose the bluff for its dramatic view of the Kansas River. Here, below the river's long arc around Weaver Bottom and the mouth of Stranger Creek, sands braided the Kaw into flowing fractals western and inviting. Wind blew ripples onto the water beside a tree-covered island that was probably dotted with morels in spring. The curve below Chance Cemetery held one of the last sandbars before the river swelled and deepened on the way to its reckoning with the Missouri at Kansas City.

Long ago, I'd paddled numberless miles on this river. Back then, few people ventured out onto the Kaw in muscle-propelled boats. But after finding an indestructible plastic canoe at a garage sale, I learned how to camp on sweet flows of sugar sand with shooting stars, fireflies, and slap-tail beavers for companions. When spring flows surged, I lugged the canoe south to the Marais Des Cygnes, French for "Marsh of Swans." In one dreamlike stretch, the stream entered a silent gallery of pecan trees, and a mink ran alongside me for five minutes, playing hide-and-seek before hopping onto a boulder and disappearing into the bulrush. In November, gusty winds rained ripe pecans into the boat like hailstones. The canoe became my gateway to wild places, to explore Kansas rivers that few people even thought to paddle: the Wakarusa, the Ninnescah, the Chikaskia, the Black Vermillion, the Fall River, the Marmaton, the Cottonwood, the Delaware, the Neosho, the Verdigris. They were booby-trapped with low-head dams and polluted; the waters sometimes ran chocolate-brown from erosion or a foul sea-green from agricultural nutrients. Illegal dumps and fish kills were common sights. But the chance to enter a magical realm at the next bend obsessed me. There were emerald gardens of watercress and beds of pink papershell mussels clinging to glacial boulders, baby foxes wrestling in nettle fields, and flocks of pelicans that covered sandbars like snow drift.

Now, change seemed to be afoot on these rivers. The Kaw had become a popular destination after the National Park Service designated it a Scenic Water Trail in 2012. An advocacy group that employed a paid "Riverkeeper" to bust illegal polluters and to police sand dredging operations, was taking hundreds of people out on the river each year. Similar groups had organized around rivers across the Great Plains and Midwest. Media outlets were covering topics related to safe drinking water, flood control, endangered fisheries, sand dredging, and conservation of watershed grasslands. Attitudes about heartland rivers, it seemed, were shifting, and prairies were part of the story.

Natural history biologist Steven Buback once told me that to effect environmental change on a prairie river, you had to "start at the top."[1] Buback ran the biggest grassland rewilding project in Missouri at the time. His statement stuck like a koan. The *top* of Buback's river—the Grand of northwest Missouri—was a maze of small streams that dissected some of the rarest prairies left in America. Restoring the river meant restoring prairies. Rivers and grasslands were yin and yang, a dual narrative, their fates inseparable. Standing beside Green Berry Chance's simple grave and looking out across the Kaw, I decided to search for the story of America's grasslands along their rivers and write about what I found.

Grassland rivers are rivers that drain watersheds that were primarily composed of prairies in pre-European settlement times. West of the Mississippi River, they flow through parts of Montana, Wyoming, the Dakotas, Minnesota, Iowa, Kansas, Missouri, Oklahoma, Texas, eastern Colorado, and extreme northeast New Mexico. Like prairies, they come in multiple varieties. Shallow and silt-laden, most undulate slowly across floodplains of modest gradient, creating sandbars, cottonwood islands, oxbow lakes, and side channels along the way. Some are birthed cold and fresh in Rocky Mountain snowmelt. Others expend their currents at lower elevations, ensconced in grassland basins for their entire journey. Prior to interstates and railroads and overland trails, grassland rivers were interior highways for migratory birds, mountain lions, Native peoples, *courriers des bois*, steamboats, and anybody who needed a dependable thoroughfare.

Over the last one hundred and fifty years, grassland rivers, like grasslands themselves, dropped from the cultural radar. Except for along a few well-traveled runs, many were less explored in recent decades than the grasslands themselves. Some were effectively off-limits due to feudal Midwestern river-access laws, others simply had been forgotten. While popular float streams in places like the southern Ozark Mountains attracted more than one hundred paddlers per hour, in any given year fewer than a dozen people laid eyes on tens of thousands of miles of grassland rivers.

Before the prairie sods were broken and the buffalo eradicated, grassland rivers had already suffered environmental insult and have since been channelized, straightened, rerouted, leveed, and dammed; poisoned from slaughterhouses, human sewage, agricultural and industrial chemicals; infested with invasive mussels and jumping carp that knock anglers from boats; and sheared of their protective forests. The grasslands were sacrificed to become America's breadbasket and their rivers, many considered sacred, its latrines. If not for their sheer hydraulic dynamism, we would have plowed them under like the bluestem itself.

The story of America's grassland rivers is larger than any one single river. With fifty thousand rivers that went by two hundred thousand names—in English, Spanish, French, Algonquian, Lakota, Kaw, Naishan, Apsáalooke, Siksiká, Chiwere, and other tongues long forgotten—distilling a route to explore them seemed daunting. Libraries were filled with books about the Colorado and the rivers of the desert Southwest; the Columbia and its salmon tributaries in the Pacific Northwest; the Rio Grande as it gathers snowmelt from the Sangre de Cristo on its way to the Gulf of Mexico; the empire-building Mississippi and Ohio and Tennessee Rivers; whitewater streams like the Chattooga in southern Appalachia; southern blackwaters like the Atchafalaya, the Pearl, and the Alabama, haunted by ghosts of the ivory-billed woodpecker; Yankee rivers such as the Hudson, the Charles, and the Concord and Merrimack, where Henry David Thoreau spent a famous week rowing with his brother. But the literature on grassland rivers was scant.

One thing seemed clear: to learn their stories, I could not be faithful to one river. One months-long, job-killing journey from meltwater to Mississippi, along the Arkansas or the Missouri, would only narrow my lens, not sharpen my focus. Mine would be a serial monogamy of rivers, a road trip into the heart of America with intense furloughs of paddling. A long float with intermittent portages of driving and work and family life. In a few cases, backpacking would be required, for reaches of the arid plains too shallow to paddle.

In the following weeks, I mouse-clicked through thousands of river miles on Google Earth, spent weekends at a university library carrel studying old maps and books stamped with checkout dates from the pre-disco era, and spoke with river guides, outfitters, kayaking companies, geographers, and prairie aficionados. I settled on an octet of clean, beautiful, and still untouched sections of rivers that scored high marks for grassland quality, scenery, recreation, ecotourism and impact on local communities, rich animal and plant life, advocacy, history, and culture. Some of the rivers were in remote outposts of the plains, where the prairies hadn't been destroyed or were making comebacks. Others were closer to home.

I set out to chronicle a type of American river whose story had never been told. Like the author William Least Heat-Moon, who searched for the "real" America by driving roads that appeared as blue lines on old highway maps, I left to find the story of the prairie and its rivers by canoe, because when it comes to exploring wild places, even the brownest river is bluer than the bluest highway. What I discovered is that across the Great Plains and the Midwest, a remarkable movement is underway at the intersection of grasslands and rivers—nothing less than a grassland river revival. This book tells the stories of these rivers of the grass.

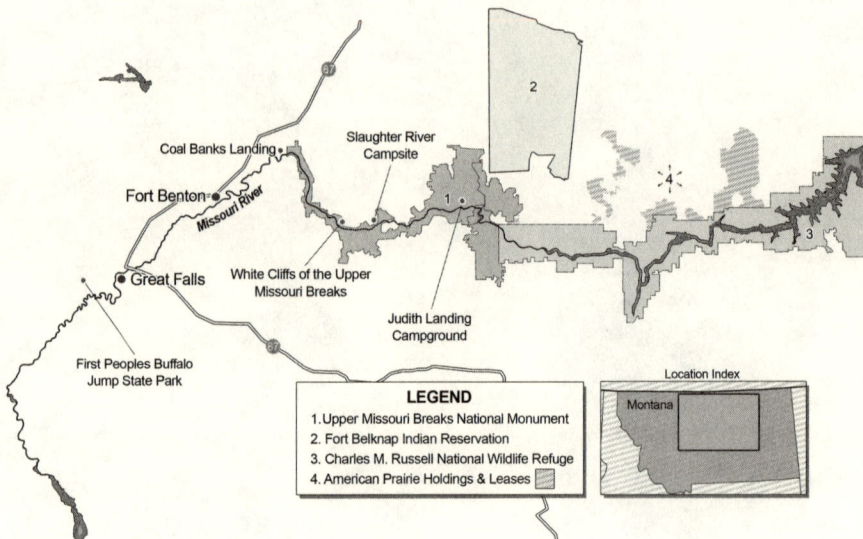

Coal Banks Landing

Slaughter River
Campsite

Fort Benton

Missouri River

2

4

1

3

White Cliffs of the Upper
Missouri Breaks

Great Falls

Judith Landing
Campground

First Peoples Buffalo
Jump State Park

LEGEND
1. Upper Missouri Breaks National Monument
2. Fort Belknap Indian Reservation
3. Charles M. Russell National Wildlife Refuge
4. American Prairie Holdings & Leases

Location Index

Montana

MAP 2 The Upper Missouri River in Montana

CHAPTER 2

MISSOURI

The river was out there somewhere in the grasses, beyond the flat yellow land that seemed to stretch to eternity. From the edge of a rocky outcrop where hunters once ran bison over the edge of a cliff, I glassed the horizon with my binoculars but could not see the actual channel, where the water ran brown between the banks. It had been out there for millennia, since before the time when glaciers flowed south, leaving breadcrumb trails of sapphires sparkling in the alluvium so geologists might one day retrace its ancient whereabouts. But from my unadorned vantage, not far from where it leaves the Rockies to become the greatest grassland river in America—in the world—the Missouri hid down in its breaks, swaddled in prairies that would sustain it for two thousand miles.

Saturday night was rising across north-central Montana. The main entrance to First Peoples Buffalo Jump State Park is near the small community of Ulm. This westernmost station of the high plains, one hundred miles south of the Canadian border, is close to Upper Missouri Breaks National Monument, where I planned to paddle for three days and learn about a bold plan to create a massive prairie reserve unlike any grassland restoration project in American history.

I hiked out onto a green of wheatgrass and blue grama so neatly cropped that it looked like the trails between the prairie dog burrows—trod by prairie dogs and perhaps a light-footed kit fox—were part of a miniature golf course. There was nothing light-footed about the prairie dogs, however. They barely moved as I passed through the colony, making me wonder if any were alive. One stuffed its root-fattened hulk down a tunnel when I got close enough to see fleas doing gymnastics in its fur. A few of the blood suckers peeled off the main swarm and bit into my legs, raising instant welts. The World Health Organization tracks plague—the illness that wiped out nearly half the population of Europe in the Middle Ages—as a reemerging disease. The bacterium that causes it can spread from prairie dogs to humans through the bites of infected fleas. I picked up the pace.

At the edge of the cliff, I dangled my legs over the drop. Thousands of buffalo had flung themselves from this spot to their misery, chased by hunters who wielded their knowledge of topography as a weapon. Hundreds of similar cliffs across the Great Plains were used this way. The *pishkun*, or "deep blood kettle" in the Siksiká language, had been used for at least a millennium. A group of hunters would run behind the herd, hopefully (for the hunters) creating a frenzy just as the bison reached the drop-off. Others would wait with clubs at the bottom of the cliff to finish the job and butcher the animals. By the time plains peoples acquired horses and stopped using the pishkun, bison bones were piled fifteen feet high. In the 1950s, a rancher took a lease on the prairie for the sole purpose of keeping it off-limits to artifact and bone pilferers. A state monument was created in the 1970s and expanded to almost two thousand acres in the 1990s, further protecting the pishkun and surrounding prairie, sacred ground to the buffalo Nations who used it. Looking west at the mesas and geological upthrusts washed red by the waning prairie light, the absence of the shaggy brutes felt palpable.

The only truly wild, never-domesticated herd of plains bison (*Bison bison bison*) ranges across the high plateau of Yellowstone National Park. Reduced to fewer than a hundred animals in the early twentieth century, the herd varies in size today between three and six thousand. Outside of Yellowstone, smaller, reintroduced but free-ranging herds of buffalo live on a handful of public reserves in the intermountain West. But the grasslands of the Great Plains are the true dominion of the buffalo, North America's great sacred ungulate, and no grassland revival on the plains will ever be complete without them.

Below its highest reaches near Brower's Spring, the Missouri River officially begins at Three Forks, where the Jefferson, Gallatin, and Madison Rivers converge. It spends a brief adolescence as a fast-moving inner-tube river, with nurseries of cutthroat trout and boulder-strewn whitewater gardens, until it reaches Canyon Ferry Lake. Soon after, and not far from the pishkun where I sat, it plunges from the mountains—wide, brown, and honest—into the lightly peopled nothingness of the Great Plains. No more a river of mountains, it becomes a river written into the history of the nation.

For years I had wanted to paddle into the heart of the Upper Missouri Breaks, a high plains prairie badlands between Fort Benton and the Charles M. Russell National Wildlife Refuge, where it feels like the hammer of time missed the spike when it was swung. The bluffs and monument rocks evoke a fabulous solitude, and for such a storied stretch of river, the breaks still receive surprisingly few recreational paddlers. This is in part because you have to float at minimum fifty miles self-supported and not be bothered by rattlesnakes. The Lewis and Clark expedition passed through in 1805. It was inspiration to Swiss painter Karl Bodmer, who documented the landscape and its indigenous peoples during an 1833 expedition with German explorer and naturalist Maximilian, prince of Wied-Neuwied. People come to listen for heartbeats of the past. The breaks have been chronicled for more than two hundred years by European and American explorers and paddlers, and by the Niitsitapi (Blackfeet) and other peoples for centuries more.

Though I appreciated those keen on finding the precise campsites of Lewis and Clark or holding up a smartphone photo of a Karl Bodmer painting at White Cliffs to marvel at the near photographic equivalence across centuries, I was more interested in what the *future* might hold for this river and its grasslands. The breaks are part of an innovative and controversial plan to build a 3.2-million-acre "American Prairie Reserve" that could one day host the complete Great Plains bestiary, including prairie dogs, antelope, deer, elk, bison, and the predators that depend on them. And, to search for pristine stretches of ordinary rivers, I'd need to see a pristine one first. Or close to it.

Happy to bask in the intense summer heat of the plains, I was excited about launching a canoe for three days in the Big Empty, but also a little scared. The route was remote and recent sightings seemed to indicate that grizzly bears, missing from the Great Plains since the early 1800s, had begun to venture back into their historic grassland dominion not far from the breaks. "It's nothing but jitters," I told myself later that night lying awake in my hotel room, listening to thunder from a storm that lit up the sky until daybreak. Despite years of experience on prairie rivers, those jitters made me question whether I was prepared for what the river might bring.

In America, no state besides Alaska is more synonymous with *wild* than Montana, but most images of its wild places are from the western high country: mountainous tracts like the Bob Marshall Wilderness, the Great Bear Wilderness, the Bitterroot Range, Glacier National Park. The eastern two-thirds of the state, however, was once covered in grasslands. On the drive from Great Falls to begin my float at Coal Banks Landing, I passed some of the largest unbroken wheat fields I'd ever seen—and I'm from Kansas. How "wild" could any river in such a farmed landscape be? But the cultivations soon withered, and near Fort Benton and later Coal Banks, the crop table abruptly collapsed into a desiccated badlands hundreds of feet below the surrounding plateau. It felt like I'd driven from the smartphone age back to the Pleistocene.

Jim, a seasonal camp host at Coal Banks Landing, scrutinized my four gallons of drinking water and chemical toilet like a TSA agent. Rules were rules. Nobody was allowed into the monument without a chemical toilet and at least one gallon of drinking water per day. The remote Bureau of Land Management (BLM) campsite was about seventy miles from Great Falls via US Route 87. As federal and state lawmakers have slashed park budgets, agencies like the BLM have turned to volunteers to maintain campsites like the one at Coal Banks. Some, like Jim and his wife, lead a nomadic existence, living out of RVs and moving among public lands in Arizona and Nevada in winter and more northern climates in summer. Volunteers aren't paid but receive food stipends and use of the campground's facilities for their RVs.

Earlier in the morning I'd rented a tandem canoe from Missouri River Outfitters in Fort Benton. Tandems are designed for two paddlers, with a seat in the rear or stern of the boat, and another in the front or bow. Soloing a tandem can be tricky. Jim helped hold it steady while I packed camping gear, food, and all my water into the front of the canoe before taking up my paddling position in the stern seat. When I sat down and wiggled a few times to get comfortable, Jim shook his head and said, "Good luck; you should have rented a solo," before heading back to the cabin.

He was right, the boat was unbalanced. The only solos the outfitter had were kayaks. I'm a canoe person. Poling the heavy Royalex jalopy forward with my paddle, I pushed the rig through calm slack water parallel to a small inlet of muddy sand. "Drink more beer to prepare next time," I thought to myself. My body weight was no match for the wind that lifted the bow and spun me around so I was facing upstream. Backward. Repeating the process three times, the wind kept slapping me back. Paddling from the stern was not going to work; it was like seesaw solitaire.

Remembering a trick from my friend Patrick Dobson, a writer who once walked from Kansas City to Fort Benton and then paddled back home, I tossed everything out of the boat and turned the canoe around so the stern faced forward. Now the "front" was just behind the yoke two feet from the middle point of the boat. After repacking the canoe and using a spare life jacket to level the seat, I paddled again toward the inlet. Minnows leaped alongside the boat like a school of joyful toy dolphins, casting a silver glimmer as the canoe cut through the wind, cleared the inlet, and joined the Missouri. The current was *moving*. At this pace I could reach Judith Landing, fifty miles downstream, in a day.

The river split into three braids near a horse pasture with a

wooden fence. In the benchland above the river I noticed movement and watched seven pronghorn antelope through my binoculars. Pronghorns are scarce in parts of the southern Great Plains,
but in Wyoming and Montana, large herds still congregate, including one that makes an annual 150-mile migration across Wyoming
known as the "Path of the Pronghorn," the longest large mammal
migration in the contiguous United States.

Established in 2001, Upper Missouri Breaks National Monument
protects 149 miles of the Missouri River that are unique among
grassland river stretches, but the river has multiple designations
within the monument. The National Park System maintains a rivers
inventory of more than 3,200 "free-flowing river segments that are
believed to possess one or more 'outstandingly remarkable' natural or cultural values."[1] Of these segments, about two hundred are
part of the Wild and Scenic Rivers System, designated for their wild,
scenic, or recreational qualities. According to the Wild and Scenic
Rivers Act of 1968, the *wild* rivers on the list represent "vestiges of
primitive America."[2] The section of Missouri that runs adjacent to
Upper Missouri Breaks National Monument is not only one of those
wild rivers, but, along with two other grassland streams—the North
Platte River at Scotts Bluff National Monument and the Niobrara
River at Agate Fossil Beds National Monument—it also runs through
a national monument. Only Upper Missouri Breaks National Monument, 378,000 acres of federal lands mostly managed by the BLM,
was created to preserve the natural and cultural history of a grassland river.

At noon I pulled over in a stand of cottonwoods at the "Little
Sandy" boat camp to cool off. The rank scent of sage was over-
whelming. Embers of a campfire still smoldered in a metal fire ring,
meaning I might catch up with other boats later in the day. Near
the campsite's pit toilet, a coiled prairie rattlesnake buzzed next to
a yucca. The previous day at the buffalo jump, I had heard a similar
buzzing as I approached a sign with the words: "Warning! You are
entering the realm of the Prairie Rattlesnake." Apparently aware of
its realm, a small rattlesnake had slithered from its spot beneath

the sign and toward the park's lone outdoor restroom, flattening its body to squeeze through a narrow gap under the closed door.

A trail led up into the bluffs to a circle of rocks twelve feet in diameter—a stone tipi ring. The rocks were hard-packed in soil. On a bluff in Kansas along the Smoky Hill, another great buffalo river, I'd once found a similar circle.

I spread out my tarp next to the ring and sat baking in the suffocating heat. All around, bands of ice-blue sage grew between tufts of wine-colored wheatgrass. From high above the channel, the fast-moving Missouri appeared stationary, a brown-skinned serpent patterned with cloud shadows waiting at the bottom of the breaks. Game trails wound through the chalky bluffs and yucca dotted the landscape like a pointillistic painting. The silence, ancient and moody, seemed as integral to the panorama as rocks, river, and sky. I wasn't completely alone, however. Near a patch of buffalo gourd, two prairie rattlesnakes buzzed their honeycombs, sounding more like cicadas than rattles. The night before at a Great Falls bar called Clark and Lewie's, I had talked to a woman who wore a rattle strung on a black leather necklace.

Botanists classify prairies into three types: tallgrass, mixed-grass, and shortgrass. These prairies occupy three vertical stripes across the central US, with tallgrass generally to the east in areas experiencing the most rainfall, shortgrass generally to the west in areas with the least rainfall, and mixed-grass in between. There are no fixed boundaries or hard definitions though, and many subdivisions exist within each type, such as the Sandhills mixed-grass prairie of northern Nebraska. Despite its placement in the far west of the Great Plains, much of the Montana prairie is of the mixed-grass type due to cooler summers and higher rainfall than regions with shortgrass prairie. Grasses dominate the flora of mixed-grass prairies and, in Montana, tend to grow in clumps as bunchgrass with prolific root systems. Western wheatgrass is the most com-

FIGURE 2.1 Eagle nest at Upper Missouri Breaks National Monument. Photo by the author.

mon species (but *not* the source of the wheatgrass juice in health food stores), along with thickspike wheatgrass, green needlegrass, and blue grama, a plant that produces a sickle-shaped cluster of flowers that bleach to camel-hide in late summer. Prairie plant communities are not monolithic. Even in micro-habitats they vary widely by soil conditions and water drainage characteristics. Sand-sage prairies are interspersed throughout the Montana mixed-grass wherever soils are sandier, such as along the washes of rivers and streams. In sand prairies, needle-and-thread grass is a dominant species, along with little bluestem, threadleaf sedge, sand blue-stem, and even the classic tallgrass species, big bluestem. But the most recognizable inhabitant of the sand-sage is small soapweed yucca—*Yucca glauca*. It produced a respectable lather of suds when my daughter and I once tried to make soap from the peeled roots.

Thousands of yuccas grew in the uplands along the river below Little Sandy, where the Missouri swept right and turned to the southeast. Now the wind was a relief from the heat, but I kept dipping a blue bandanna into the water and draping it over my neck to cool off. On one island, a bald eagle perched on a lone cottonwood, indifferent as I paddled close. The eagle served as the undesignated sentry to the official "wild and scenic" section of the river. This section is open only to nonmotorized traffic from June 15 to September 15, marking the beginning of the White Cliffs, one of the most dramatic sections of any flatwater river in America.

The Upper Missouri Breaks region has been inhabited for at least 12,000 years. At the time of first European contact, the Nakota (Assiniboine), Niitsitapi (Blackfeet), A'ani (Gros Ventre), Apsáalooke (Crow), Néhinaw (Cree), and Anishinaabeg (Plains Ojibwa) inhabited the area. Other peoples visited from the eastern plains to hunt. The Lewis and Clark Corps of Discovery expedition traveled upstream through the breaks in May and June of 1805 and passed through the White Cliffs on May 31. Meriwether Lewis and a small party returned on their way back east in July 1806, while William Clark and others explored an alternate route down the Yellowstone, one that joins the Missouri in present-day North Dakota west of Williston. The locations of ten Corps of Discovery campsites are known in the national monument, part of the Lewis and Clark National Historic Trail. Visitors come to camp at those sites and compare the landscape to descriptions from the journals. Of the White Cliffs Lewis wrote: "The bluffs of the river rise to a hight [sic] of from 2 to 300 feet and in most places nearly perpendicular; they are formed of remarkable white sandstone which is sufficiently soft to give way readily to the impression of water."[3]

Tales of Lewis and Clark continue to cast a long shadow on the history of the Missouri because the narrative nature of their journals told a reader-friendly story of the expedition. The writings have inspired musicals and melodramas with scintillating taglines: "Will Lewis and Clark survive their journey and become American heroes, or will it all just be a waste of time?"[4] The travel logs of other early white explorers were less suited to dramatic interpretation, but some contained different perspectives from Lewis and Clark's. The travels of Maximilian of Wied-Neuwied in 1832–33 paint a different picture of the Missouri in part because of his choice of media.[5] Maximilian, a science-educated German prince of the European enlightenment, had traveled through South America fifteen years earlier. Contemporaries criticized the poor quality of the self-drawn illustrations in his subsequent book.

So when he turned his attention to the northern plains, Maximilian brought along skilled Swiss painter Karl Bodmer, known for his precision landscapes in the pre-photographic era. The aquatints made from Bodmer's drawings are famous today, and his most encyclopedic work was done in the White Cliffs. Less frontier-style "explorers" than ethnographers/travelers studying indigenous cultures from booked passages on American Fur Company steamboats, Maximilian and Bodmer documented their encounters with the Numakaki (Mandan) and Hiraacá (Hidatsa) peoples prior to severe smallpox epidemics that swept through the Nations of the upper Missouri, decimating Native populations. Unlike Lewis and Clark, who brought an enslaved man named York to handle the most back-breaking jobs, Maximilian was outraged by the practice of slavery and the treatment of Native peoples, as he wrote, "in this land of vaunted liberty!"[6] The names of many of the formations in the White Cliffs are recorded in Maximilian's journals and serve as the basis of modern guidebooks. The precision of Bodmer's art allows paddlers to compare the 1830s White Cliffs to their modern incarnation in a way that's impossible on any other grassland river.

The fur trade industry spawned extensive travel along the Missouri, by keelboat, mackinaw (a lightweight open-topped wooden boat with a short sail), bullboat (made from willow in the shape of a circular upturned bowl), and canoe. The first steamboat arrived in 1831. Steamboat traffic increased after gold was discovered at Grasshopper Creek, triggering the Montana gold rush in 1862. American settlers followed after the Civil War, and some tried farming the Missouri bottoms. Ruins of their shacks and livestock enclosures in the national monument are testimony to the poor conditions for agriculture in the years before the Dust Bowl.

As my canoe hugged the current down the center of the river, the cliffs stood three hundred feet high in spots on either side of the channel. Lewis described them as "lofty freestone buildings having their parapets well stocked with statuary."[7] Vertical gates of dark caramel-colored shonkinite, a rare volcanic rock found only

FIGURE 2.2 Shonkinite wall at Upper Missouri Breaks National Monument. Photo by the author.

in Montana, Ontario, and on the southeast Indonesian island of Timor, cantilevered between cliffs in what looked like constructed walls, perhaps fences for dinosaurs. Millions of years ago molten lava forced its way up through layers in the sediment to form sheets that have since eroded to expose sandstone sandwiches of lava. Burned Butte rose high on the left, a torqued lava flow that looked like a whirlwind frozen in rock. An osprey flew by clutching a fish and landed on one of the pearly walls before taking off again for a nest downriver. Osprey are summer residents, but the scorched environment seemed a strange home for this "sea hawk" of coastal estuaries and salt marshes.

I still hadn't seen another boat when I reached Eagle Creek campground, one of the most popular camp spots on the upper Missouri River. The author and historian Dan Flores, in his quintessential treatise on Great Plains fauna *American Serengeti*,[8] described Eagle Creek as the single greatest spot to wake up in your tent to a vision of what the pre-European settlement plains might have looked like.

I didn't camp, but continued to paddle past Grand Natural Wall; Neat Coulee; LaBarge Rock; the remnants of Eye of the Needle,

FIGURE 2.3 BLM Hole-in-the-Rock campsite in the White Cliffs section of Upper Missouri Breaks National Monument. Photo by the author.

which collapsed in the 1990s; Eagle Rock, near the first of three golden eagle nests on my route; and finally, about twenty miles below Coal Banks Landing, Citadel Rock, one of the most famous landmarks on the Missouri River, an igneous remnant of weathered lava that looks like a carving of a person in a robe. It cast a perfect and unnerving reflection in the river. I had to close my eyes paddling through it to ward off vertigo.

At Hole-in-the-Wall boat camp, I finally encountered people. The riverside meadow ran brim to brim with Boy Scouts from Great Falls and Fort Collins. I beached my rig on a low mudflat, and two of the kids helped me ferry bags onto shore. After changing into shorts and a T-shirt, I looped the bow rope a few times around my shoulder and stepped the canoe back into deeper flow to guide it downstream to a better spot for launching the next morning. The fast current and its heavy load of sediment pushed against my legs and torso. The cold water felt good after thirty miles of paddling.

The BLM campsites were well maintained for such a remote place. This one had a vault toilet, metal fire rings, and two open-faced log shelters that would be useful in heavy weather. Fortunately, although a thin overcast would veil one of the darkest night skies in the United States, no storms were on the horizon. I pitched the tent near some green ash saplings.

It was strange to be back among people after a day alone with the osprey, golden eagles, ravens, magpies, antelope, and a small herd of elk at the base of the cliffs near the end of the float. Later that night, another group landed, and four whispering people set up their tents around mine. They spoke of storms, and at 10:30 p.m., peals of thunder woke me. I rolled back over in my sleeping bag. One of the men said, "I'm not worried, they're way off, this is the only thunder we'll hear all night."

The woman with him said, "You don't think it will be like last time?"

"I hope not."

For most of American history, grasslands as wild places have been invisible. Maybe they always will be. We equate heartland with substance and sustenance, an energy- and mineral-producing grazeable breadbasket that provided a growing nation with three squares a day. But heartland as wilderness? Native grasslands were nearly obliterated after European explorers and settlers began arriving. By the end of the nineteenth century, not only were the great herds of bison, antelope, and elk gone but also the predators that relied on them. Most of the grasslands themselves were plowed, destroying habitat for prairie dogs and their dependents such as swift foxes, burrowing owls, and black-footed ferrets.

This invisibility extended throughout the first wave of national-park building. Between 1875 and 1925, the federal government enshrined ecosystem-scale sanctuaries in Yellowstone and Glacier National Parks and conserved tens of millions of acres of heritage landscapes and biomes across the American West. Mountains and deserts became the conservation focus, in part because their scenic virtues didn't conflict as much with their mineral and agricultural potential. Park planners skipped the sod lands, not only in the farmed-over tall prairie regions of the Midwest, but also in the Great Plains where bison and their cohorts had thrived not long before. Especially in the arid high plains, rewilding would require real estate, a landscape to scale with space for roaming. Not mountaintops.

In the twentieth century, state park systems and private groups like the Nature Conservancy began preserving prairies through direct purchases and conservation easements, including the 39,000-acre Joseph T. Williams Tallgrass Prairie Preserve near the Osage Nation in Oklahoma, Missouri's Prairie State Park, and Konza Prairie in Kansas. At the federal level, President Theodore Roosevelt designated the 60,000-acre Wichita Mountains Wildlife Refuge in southwest Oklahoma in 1905. Federal legislation aimed at stabilizing soils at the end of the Dust Bowl led to the creation of twenty national grasslands managed for "many uses" by the National Forest Service. Elsewhere in parts of the plains with a high percentage of private property, the lack of a budget for land acquisition was one difficulty in establishing parks administered by the National Park Service (this changed in 1964 with passage of the Land and Water Conservation Fund Act). Eventually, Tallgrass Prairie National Preserve in the Kansas Flint Hills, White Horse Hill National Game Preserve in the Nebraska Sandhills, Wind Cave and Badlands National Parks in South Dakota, and other smaller preserves were established to protect grasslands, several with managed bison herds. Although America had nothing like the world biosphere parks and game reserves of the Serengeti grassland in northern Tanzania, people have dreamed of massive Great Plains parks ever since the bison were hunted almost to extinction.

In 1999, the Nature Conservancy published an exhaustive inventory of large grasslands in the northern Great Plains with sufficient biodiversity and never-plowed prairie to viably preserve what remained of the once vast "Northern Great Plains Steppe."[9] The World Wildlife Fund, an international conservation group known for preserving African lakes and Central American rainforests, began to explore rewilding possibilities in Montana near the Upper Missouri Breaks, one of the areas identified in the study.

This wasn't the first time a nongovernmental entity had proposed rewilding the Great Plains. When he was a college student, Chicago native Frank Popper became fascinated with sparsely populated landscapes like the Great Basin in Nevada and the Great Plains while driving through them on summer road trips to California. After he graduated and started working as an urban planning professor at Rutgers University, Popper focused his work on the least populated metropoles in America—places with accelerating *depopulation*. In the late 1980s, the Great Plains epitomized this trend. Excluding cities like Denver and Oklahoma City on its periphery, the entire region—one-sixth of the American landmass—had less than six million people. He and his wife, Deborah, a geography graduate student, mapped census trends for plains counties whose populations reached their peaks between 1890 and the Dust Bowl, before sliding into population decline due to weak farm and ranch economies and a general migration from rural to urban areas. In some counties, the median age bumped against sixty years.

Their work, published in an obscure planning journal, created a sensation.[10] Not because of their data—population decline on the plains was well known and there were ghost towns to prove it—but because of their proposed solution. Eventually the plains would lose so many people that the federal government could buy back land that it had "given away" during homesteading (after stealing it from Native Americans) and create a 90-million-acre "buffalo commons" in ten states where the pre-European settlement conditions of the Great Plains could be revived.

Political leaders in the heartland scoffed. It didn't help that the Poppers were academics from New Jersey. In Kansas, some suggested that the government should buy up land in New Jersey and convert it to a national pig farm because the idea was hogwash. But the trends the Poppers predicted have continued. Thirty years of new technology have not reversed population decline in the high plains.

In the third decade of the twenty-first century, a ten-state preserve remains a fantastical notion. There is zero political interest in reviving legislation like the Bankhead-Jones Farm Tenant Act of

1937, which authorized the federal government to buy up ravaged farms during the Dust Bowl era. But the Nature Conservancy and World Wildlife Fund used new science to establish a lower boundary on the magnitude that any self-sustaining "buffalo commons" ecosystem would require: 3.2 million acres. This was much smaller than the Poppers'· proposal, but enormous nonetheless, on the scale of world biosphere preserves in Africa and South America. The grasslands in and around Charles M. Russell National Wildlife Refuge (CMR) and the Upper Missouri Breaks National Monument were the best candidates, and the CMR already constituted one-third of the requirement at 1.1 million acres. Experts at World Wildlife Fund decided that such an ambitious project needed a new, dedicated organization to have a shot at success.

That organization—now called American Prairie—would take a decidedly different approach to park-building, using a model from a place that seemed antithetical to environmentalism and ranching culture: Silicon Valley. Its leaders developed a disruptive plan with a new vision for conservation in a landscape that had thwarted previous attempts at park-building, aimed at creating a preserve that would be bigger than Yellowstone and Glacier National Parks combined. American Prairie would build the reserve without a single acre of state or federal land acquisition, a policy that, in the political climate of twenty-first century America, seemed to give the project a better chance to succeed. At least at first.

To understand the motivation behind American Prairie, I spoke with Daniel Kinka, senior wildlife restoration manager for the organization. In graduate school, Kinka's research focused on ways large predators and humans can coexist without killing each other.[11] Since the technological advances of recent centuries heavily favor the latter (try to remember the last time a candidate campaigned on a "right to arm bears" platform), this usually involves reducing interactions between wild carnivores and ranch animals. He studied how "livestock guardian dogs," like the ones that herd sheep, could help reduce attacks by wolves and grizzly bears. Kinka's academic background in coexistence is well attuned to the general philosophy that helped launch the American Prairie Reserve.

The organization formed in 2001, after Sean Gerrity, cofounder of a management training and development company in Santa Cruz, California, met Curt Freese of the World Wildlife Fund shortly after Gerrity moved back to his hometown of Great Falls. Freese was an ecologist who studied pragmatic solutions to big problems; he specialized in the intersection of economics and biodiversity preservation. Gerrity trained managers of startups to navigate technological ecosystems. He was a Montana guy at heart and loved the open range. Along with a core group of cofounders, Gerrity and Freese crafted a vision for the reserve that would wed a free-market approach to fundraising with a radical land acquisition model that had never been tried on such a large scale. Gerrity became the first president of the fledgling organization. Their goal was to create the largest wildlife preserve ever dreamed of in North America, one that, when fully realized, would exceed three million acres managed completely for biodiversity.

Their conservation model was disruptive, in the Silicon Valley sense of the term. The reserve would never become a national park and would not depend on government land buyouts. Instead, American Prairie would purchase land from willing sellers one parcel at a time to stitch together a reserve of high-quality native grasslands like a quilt, with the 375,000-acre Upper Missouri Breaks National Monument and the 1.1-million-acre CMR as the foundation. The reserve also would target nearby BLM and Montana state lands through grazing leases. American Prairie would hold title to the lands it bought, including conservation easements and grazing rights on public lands.

This was not a mere academic exercise. In a twenty-year period, they raised more than $100 million, purchased 117,000 acres, and attained initial leases on another 335,000 acres of state and federal public lands. That's an enormous chunk of real estate.

Kinka explained, "American Prairie acquires private lands—ranches—to stitch public lands together, utilizing them a little bit differently than a cattle producer, i.e., grazing bison. We serve as caulking. Even though it will always be collaboratively managed, we have no vision to turn over or deed our properties to any kind of government entity and we can do this without government subsidies, government money, or government designation. We will remain a private partner, NGO to these other public lands in perpetuity."[12]

The model is controversial and has faced resistance in Montana. The assumption that biodiversity will be better served by American Prairie than by cattle ranches has become a source of conflict. Ninety percent of the land targeted by American Prairie has never been plowed, in part because of the dry climate that makes it difficult to grow row crops and in part because Montana ranchers have not only preserved grasslands but also restored overgrazed prairies and introduced modern conservation management like rotational grazing, regenerative agriculture, and range burning.

Kinka said that American Prairie doesn't dispute the importance of responsible ranching. They are trying to work with local ranchers to develop economic synergies between wildlife and the ranches. One program, Kinka said, is called Wild Sky. "We've known from Yellowstone, for example, that hard boundaries around wildlife preserves don't work very well. At 3.2 million acres of land managed for wildlife, if there is not tolerance outside of the reserve, you could wind up with a kind of demilitarized zone at the edges. American Prairie's neighbors are always going to be ranches, now, in twenty years, in a hundred years. If there is little tolerance for wildlife abundance at the perimeter, there will be less wildlife abundance within the American Prairie Reserve."

Wild Sky pays a reverse bounty on wildlife. Instead of rewarding ranchers to kill predators, it pays ranchers to photograph them. Kinka explained: "Cattle ranches provide a considerable amount of public good by providing habitat for public wildlife; the ranch owner is not compensated in any way for the services that they provide. Rather than pay producers for animals that get killed by predators or grass that elk eat, what if there is a proactive approach where we value and monetize habitat for wildlife? We collaborate with ranches to set up game cameras and pay people for the pictures that they take. Every picture of a coyote earns the rancher a payment, right now about $25. It would be hundreds of dollars if a grizzly bear or wolf ever showed up. An easy way to think about it is paying rent for animals."

I spoke with Wild Sky rancher David Crasco, who was initially opposed to the program. "I was one of the naysayers," he told me. "But a neighbor of mine had worked with them. I was already beginning to install wildlife-friendly fencing on my own, where the bottom wire is smooth and eighteen inches off the ground, and the top wire is low enough for a deer or elk to hop over. Since I've joined Wild Sky, they provide bales of the wire, although I do the work. Since I've switched to wildlife fencing, the maintenance costs on my fences have gone down considerably."[13]

Crasco's ranch is on the Fort Belknap Reservation. A fourth-generation cattle rancher, he's a member of the Nakota Nation and served on the Fort Belknap Tribal Council. He said, "To be a good councilman, you need to approach things with an open mind and hear all sides of every issue or you're going to make a poor decision. Despite my initial opposition, I listened to what they were doing. Now I've been with the program for six years."

Camera traps on Crasco's ranch have captured photos of bears, coyotes, mountain lions, wild turkeys, and elk. He's been paid for each shot. "The money is not going to make or break me, but everything helps," he pointed out.

Rewilding is at the core of American Prairie's land management strategy. They use metrics developed specifically to measure grassland biodiversity to guide the work. The "Freese Scale," developed by their cofounder, tracks ten ecological management factors for temperate grasslands, including soil and vegetation management, the presence or absence of seasonal fire, hydrology, ungulate populations, big predators, and habitat fragmentation. Biologists run the numbers on each American Prairie parcel and develop land management practices based on the results.

Compared to the Midwest and most of the Great Plains, rewilding grasslands at American Prairie benefits from the near-pristine condition of the range. Kinka said that commodity cycles inject urgency into their mission: "This part of Montana has never been a good place for row crops, but you see pulses where land gets plowed up when grain prices get high. People will give it a go and usually fold a couple of years later. But then the damage is done. We're in a race against the next boom. Once plowed, it will take fifty to a hundred years to reestablish a biodiverse mix of grasses and forbs with a complex array of soil microbiota and complicated root communities that nobody, I think, really understands. We want to get this thing done before we lose any more of that virgin prairie."

American Prairie's Montana concept of "rewilding" differs from other prairie rewilding projects in one key manner. Rather than replanting grasslands, Kinka said, "to us, rewilding means megafaunal rewilding. The timelines to reintroduce bison and other large grazers are much shorter than prairie rewilding. I could see it completed in my lifetime, and easily in my daughter's lifetime. It only works if you're focusing on lands that are relatively intact to start with." In 2005, they reintroduced the first bison on the reserve. As of 2024, the herd numbers about eight hundred. In accordance with Montana law, bison are treated like livestock rather than wildlife. They receive veterinary checkups and inoculations to ensure none is infected with brucellosis or other diseases that can spread to cattle. Someday the reserve aims to support ten thousand bison, potentially the largest free-ranging population in the world.

One of the fallacies of the term "wilderness" is that it is unpeopled. The Wilderness Act of 1964 defines "wilderness" as "an area where the earth and community of life are untrammeled by man, where man himself is a visitor who does not remain."[14] Using that definition, Antarctica would have been the only continent with wilderness for most of human history. The North American grasslands have been peopled for millennia. For American Prairie to succeed, it needs to coexist with ranching communities, Native American Nations, and their economic activities. Kinka said that eventually the fence around the reserve will become meaningless to the bison: "Our job is to create such a huge interior that most bison will be born and die without ever seeing a fence."

But not everyone in Montana agrees that the preserve can coexist with cattle culture, no matter how big a fence it builds.

When I woke before sunrise, the ground was still dusty, the night's rain nothing but mist. Jackrabbits scattered when I crawled out of the tent. One of the scouts had waded into the river and waved as I packed my gear for the morning. Western meadowlarks were calling in the yellow grass and a flock of dentine-white pelicans landed on the river. As the eastern sky lightened behind the played-out storm clouds, the chalky beige rock towers along the canyons emerged from darkness, splotched with cedars. My goal for the day was modest. Beat the swarm of scouts to Slaughter River and nab the best camp spot, selfish though that thought felt. If the mist returned, I might need to dry out my gear. Slaughter River, like Hole-in-the-Wall, had two log shelters. It would be good to rest under a roof of sorts.

The reverse seating arrangement worked well the first day, but now something was off center in the canoe, and soon after I launched, it started doing pirouettes in the main current. One of the scouts watching me depart said, "We're gonna do *that* today."

After rebalancing the gear, the river did all the work while I enjoyed a rock concert of pillars, spires, and prairied buttes. I stopped to hike at Valley of the Walls, a badlands scene of receding canyons, juniper-dotted hills, and broken fences of shonkinite, but didn't make it far before I realized I could either spend the day hiking or keep paddling. Back in the canoe, the sun emerged and flooded the river with oblique light, illuminating little salads of underwater plants waving in the current. The channel weaved through statuary formations and historic landmarks: Dark Butte, the Archangel, more Lewis and Clark campsites, Pablo Rapids (just a ten-second patch of rough water), an old ferry crossing that I couldn't find, and Wolf Island, named not for a wolf but for a steamboat captain whose crew mutinied, sent him ashore, and drank all the whiskey on the boat. Clearing the shallow rift where the two forks reconnected below the island, I waved off a bloom of caddis flies in the still water of the eddy that followed. The landscape was silent, like the day before, except for the bellowing of cattle that came down to the river in large numbers. In places they had stomped the banks into mudholes, and their filth mingled with the river in stagnant backwaters.

It didn't take long to reach Slaughter River. Tired and disgusting with sweat and grime, I wedged the boat against a mud bank and claimed one of the log shelters. Five minutes later the Cappers family, the people who had camped next to me at Hole-in-the-Wall, arrived and claimed the other. As soon as I located my packet of dehydrated beans and fired up the stove, the first of the Boy Scouts splashed ashore. They couldn't have cared less that we beat them, but I think some of their chaperones envied our spots in the log hangouts.

Dan and Sue Cappers had paddled the breaks fifteen years earlier when a man was nearly decapitated by a windblown canoe at Judith Landing. This fact emerged as we talked across a raging campfire between the two log shelters at Slaughter River. A thunderstorm had blown in during the night as they camped with their two young children at Eagle Creek. The next morning, they met a group in voyageur canoes dressed as nineteenth-century fur trappers. Dan asked one of the reenactors if they had been in the storm the night before. The man said, "That wasn't a storm. That was the end of the world."

He'd had enough voyageuring; the violent weather had triggered his latent heart condition. He decided to take a canoe, leave the reenactors, and catch a ride back to Great Falls once he reached Judith Landing, a hundred miles short of their intended destination. When the Cappers got there two days later, the camp host told them that after the man got off the river, he had failed to tuck the nose of his canoe into the locking slot of the boat storage locker. As he walked back from the main camp, a gale roared down the canyon and dislodged his boat, turning it into an airborne missile that knocked him out. It was a miracle he wasn't beheaded.

I left the Cappers to their reunion, and noticing there were no other fire pits, decided to set up my tent in the grass and let the Boy Scouts know they could take over the log shelter. The forecast called for light rain, but thinking of the reenactor, I attached the rain fly, and was asleep as soon as my head hit the stuff sack filled with my dirty river clothes.

The upper Missouri was designated a wild and scenic river by an act of Congress in 1976. Lee Metcalf, the Montana senator who was instrumental in passing the Clean Water Act and designating large tracts of Montana as wilderness in the 1960s and 1970s, sponsored the act. In 1971, he and Utah senator Frank Moss held public hearings in Great Falls. At the meetings, dissenters argued the designation could stunt future job growth and attract the "wrong type of people." Big Sandy Rotary Club submitted a letter stating its members "cannot see the human needs of our future generations giving way now to a few selfish and short-sighted minority groups' recreational desires at the government's expense and provide more government property for 'hippies,' undesirables and other human pollution to camp on and dare you to touch them."[15]

But one critical local stakeholder came out in favor of the wild and scenic designation. The Montana Stockgrowers Association has been the primary advocacy group for Montana's cattle grazing community since 1885, when it was formed to take on cattle rustlers. Almost a century later, the group "wholeheartedly" supported the wild and scenic designation. At the public hearing, secretary Mons L. Teigen said "this area has been able to retain its wilderness character throughout the years due to . . . the fact that the primary use has been by the range cattle industry, which disturbs the natural environment about as little as any commercial industry in Montana."[16]

The wild and scenic designation emphasized a multiuse land model that blended recreation with ranching, mining, oil and gas leasing, archaeology, and historical research. It would be administered by the BLM under provisions of the Taylor Grazing Act of 1934, the law that effectively ended homesteading in the West and established rules for who could graze cattle and sheep on BLM lands. The BLM manages over 155 million acres across the West, and according to the agency, as of 2024 roughly 18,000 permits and leases are held by ranchers on 21,000 allotments.[17] Leases last for ten years and are renewable if the BLM finds the terms of the agreements are being met. Because cattle would continue to be allowed on pastures in the breaks, and the proposal further protected those lands against future water impoundment projects (no new dams would be allowed between Fort Benton and Judith Landing), the wild and scenic designation was congruent with the interests of Montana Stockgrowers Association members.

In 2017, American Prairie submitted a proposal asking the BLM to modify details of its use permit on seven grazing allotments it controlled in Phillips County. They asked to (1) change the class of livestock from cattle to "cattle and/or bison," (2) allow year-round grazing for "indigenous animals" (bison), and (3) allow construction of exterior fences and removal of interior fences between allotments on pastures that would include bison.[18] About 260,000 federal acres were involved in the request, as well as almost 30,000 acres of State of Montana land. If approved, the changes would allow American Prairie to increase the size of its bison herd from eight hundred to one thousand animals. Unlike with the 1976 wild and scenic river designation, American Prairie's 2017 proposal faced strong opposition from the Montana Stockgrowers Association, as well as from groups representing central and eastern Montana property owners, the state attorney general, and the governor.

This wasn't the first time American Prairie faced local friction. A wide consortium of ranching interests had opposed the American Prairie Reserve from its inception, and much of the resistance revolved around property rights. Deanna Robbins, who runs a cow/calf operation and big game outfitting business with her husband, Mark, in Fergus County, cofounded the United Property Owners of Montana (UPOM). The organization believes that American Prairie

Reserve's stated goal of acquiring over a million acres of private land would "wipe from the map" hundreds of family farms and ranches and fundamentally change the ag-based economy around the preserve.[19] Most of American Prairie's funding, UPOM members have argued, comes from out of state, and American Prairie uses its tax-exempt status as an NGO and "Silicon Valley" funding apparatus to unfairly bid up the price of Montana land. If a "willing seller" knows that American Prairie needs their ranch to complete part of a wildlife corridor for the reserve, the price of the land will inflate accordingly. The UPOM argues that ranchers and agricultural producers can't compete fairly with this model.

UPOM, the Montana Stockgrowers Association, and Montana Attorney General Austin Knudsen opposed American Prairie's efforts to make BLM grazing more bison-friendly from the beginning. The interpretation of the Taylor Grazing Act and subsequent laws and the question of how public land should be used are the crux of the disagreement, one that has been waged in different incarnations across public lands in the American West for decades.

The status of bison as "livestock" rather than "wildlife" is a unique point of contention. In 2013, the Montana Department of Fish, Wildlife, and Parks (FWP) considered a proposal to transfer wild bison from Yellowstone National Park to the Fort Peck Reservation and other locations in central and eastern Montana to combat overpopulation of the Yellowstone National Park herd. In the park, bison were considered "free-roaming" and under FWP's proposal would remain so after transfer. Ranchers opposed the move, arguing that free-roaming bison would destroy fences, compete for forage, and, because no entity was responsible for bison considered "wildlife," financial hardship could befall landowners if transplanted bison caused damage. The Montana Supreme Court eventually allowed the transfers, but with one major contingency: all bison would first need to be quarantined to check for brucellosis. Chief Justice Mike McGrath wrote in the decision that a "wild buffalo or bison is defined as a bison that has not been reduced to captivity and is not owned by a person."[20] The quarantine would count as captivity, so by transferring them outside of Yellowstone, bison would become livestock from a legal perspective.

Attorney General Austin Knudsen took the issue further. Based on feedback he gathered at a public listening session he hosted in Malta, Knudsen sent the BLM a letter arguing that American Prairie was essentially trying to change its BLM leases to treat bison, legally considered livestock, as wildlife. Knudsen argued that bison are not and should not be considered livestock according to current law, which reserves the term for domesticated animals. Bison, as American Prairie defined them in its proposal, were "indigenous animals," and he wrote that American Prairie's marketing materials made it clear that the organization believes bison are not livestock. The terms "indigenous animals" and "indigenous livestock," he continued, had no legal definition, and buffalo are clearly not cattle, sheep, or goats. He claimed this is important because the Taylor Grazing Act and its successors were created to provide a grazing lease structure "to provide for the sustainability of the western livestock industry and communities that are dependent upon productive, healthy public rangelands."[21]

When the BLM eventually approved American Prairie's proposal in 2022, Knudsen and Montana governor Greg Gianforte asked a federal board to overturn the permit. The Montana Stockgrowers Association also came out in opposition, writing in a press release that "bison's alleged ability to improve rangeland conditions fits more in the realm of 'urban myth' than as an established scientific fact," and that "American Prairie's American Serengeti concept is a threat to the grassland ecosystem, rural communities, and Montana's cattle industry."[22]

These disputes are nothing new. In the 1980s and early 1990s, environmental groups created a bumper-sticker-friendly campaign to make BLM lands "Cattle Free by '93." They argued that BLM range is federal property owned and maintained by the tax dollars of all US citizens. In 2023, the fee to graze a cow on publicly owned BLM land was $1.35 per month, the lowest legal limit set by a 1986 executive order signed by President Ronald Reagan. If fees had kept up with inflation since then, the current price would be greater than $7 per month. During that time, the market for US beef has sky-rocketed, from $300 million per year in exports in the nineties to over $3 billion today.[23] Does grazing cattle at below-market grazing fees serve the public interest more than grazing bison, an enigmatic American animal that once nearly vanished from the planet? There are thirty million beef cattle in the United States. American Prairie wants to graze 10,000 bison, some of them on public lands. Is there a middle ground that can serve both American Prairie and the cattle industry—especially since conservation of the range is fundamental to the success of both groups? The one constant in the history of the Great Plains over the last two centuries is change, and American Prairie is just one in a long series of newcomers. Part of its success will depend on how well it can spread that success among local communities in coming decades.

Early the next morning while the Boy Scouts were sleeping off their s'mores, I paddled across the river and tied off to a boulder on the opposite shore. Cliff swallows flitted from mud nests attached to a sheer cliff of sandstone, one of the most beautiful standing forms in the breaks. Modern Arrow Creek is just downstream from the cliff. Lewis and Clark called it Slaughter River because they found the rotting carcasses of buffalo that they believed had been driven over the edge. Lewis wrote "today we passed on the Stard. side the remains of a vast many mangled carcasses of Buffalow."[24] He thought it was a pishkun. The Niitsitapi told them later that the buffalo had drowned and piled up to form an ice jam. There was a dead smell close to where I anchored, and thinking of the rotting buffalo, I paddled toward it and found a calf carcass in a rush shallows, drowned like the bison more than two hundred years before.

Back at camp, I took my time packing but still was on the water before the Cappers or Boy Scouts began making breakfast. A half mile downriver, the current swung river-left, and I beached the boat on a gravel bar next to a prairie meadow that held a thin layer of fog in a low spot. Unhurried, I set up my camp stove and clicked the flint to brew one more cup of tea. The sun rose, bringing out the olive hue of the river as it followed the contour of white sandstone inlaid in the bluff for centuries. My presence here felt insignificant, and perhaps because of that, more real for some reason. Hording the scene felt selfish, but sipping tea, I tried to burn its features into my memory. In the Midwest I would be searching for connected rivers like this, where the prairie grew right up to the riverbanks.

Below Slaughter River, the Missouri left the White Cliffs and entered a more rugged section of badlands. Feeling stronger, my muscles settling into the rhythm of daily exertion, I zenned out on my form for the next two hours, paddling hard to Deadman Rapids where a canoe had capsized in 1837, drowning four men. The rapids hid until I was upon them in a rocky shallow. The canoe snagged and spun right. I had to stand and pull it free of the rocks. Perhaps the rapids played similar hijinks on the unfortunate men, who would have carried a much heavier burden than my meager payload.

On edge because of the tussle with the rocks, I ducked into an eddy to recalibrate. Downstream, on the same side of the river, something large swayed back and forth in the grasses close to a cottonwood tree where a golden eagle had made its nest. I froze. The motion seemed menacing. A large brown animal, bigger than a full-grown cow, stood at the base of the tree, watching me in the canoe. Quietly, I felt through the dry bag for binoculars, but thought better and grabbed the paddle instead to ferry across to the other shore.

Earlier at Coal Banks Landing, when the camp host, Jim, helped me pack my gear, he had found my canister of capsicum self-defense spray and asked me what it was for.

"Grizzly bears?" I offered, like potential prey.

He turned his head away and smiled. I added, "I made a last-second trip to a hunting store last night to buy this stuff."

"Well, I'm sorry you won't get to use it. There are no grizzly bears along this river."

Now, sitting in my canoe and looking at the dark and curious hulk, I wasn't so sure.

Kinka of American Prairie told me that, although the eventual goal of the reserve is to support the full cohort of Great Plains flora and fauna, they have no plans to reintroduce apex predators. Species reintroduction is under control of Montana Fish, Wildlife, and Parks and the US Fish and Wildlife Service. American Prairie maintains that these agencies and the will of the public will guide any future decision about big predators.

There are two species of bear in the continental United States: black bear (*Ursus americanus*) and grizzly bear (*Ursus arctos horribilus*), America's largest terrestrial predator. Prior to 1850, grizzlies ranged across most of the western United States, but by 1950 were extirpated from all but 2 percent of their former range. This included the Great Plains. The great bears followed the buffalo and, like people, preferred soft well-watered valleys and grassy plains to the high mountains. But the high mountains are where most of the estimated 1,500 grizzlies that remain in the lower forty-eight live today, in Yellowstone and Glacier National Parks, northern Washington, and a few other deep wilderness habitats in the Rocky Mountains. Almost one thousand are thought to live in Montana.

In 2009, a grizzly bear killed a sheep near Loma, north of the river between Fort Benton and Coal Banks Landing. It was trapped and released in the mountains, but the next year returned and was found foraging grain in the same area. In 2017, Montana Fish, Wildlife, and Parks rangers were called out to investigate two young grizzly males near Stanford, in the plains southeast of Great Falls. The bears were euthanized after they killed two calves. At the time, according to the agency, this was the farthest east grizzly bears had ventured since the early twentieth century. Grizzly dens have been found in the plains, one on the Blackfeet Reservation in 2009 and one near Choteau in 2012. In both cases, denning females gave birth to a pair of cubs, perhaps the first grizzly young born on American grasslands in a hundred years. In the summer of 2020, a farmer was attacked by a grizzly he found in his barn on a farm near Choteau and had to be taken to a hospital.

Kinka said that in September 2021, there was a confirmed grizzly bear track in the Upper Missouri River Breaks National Monument, within American Prairie Reserve target lands. Rebuilding bison numbers is supposed to precede the return of large predators, but nobody told the bears. Every year since the turn of the century, a few have been roaming back onto the plains, whether a bison buffet awaits them or not. Grizzlies are protected under the Endangered Species Act. Efforts are ongoing to delist them in the Yellowstone ecosystem and beyond. That could allow states to open hunting seasons. Wyoming and Idaho planned grizzly hunts in 2018 until a federal judge blocked them. Mountain populations, much less grizzlies on the Great Plains, are controversial. But grizzlies will probably keep venturing onto the grasslands, along the Missouri River and its tributaries, and at least for the foreseeable future, at their peril.

I raised my binoculars toward the motion downriver and saw a lightning-torched cottonwood trunk, an ursine mirage manufactured by the bear-fearing part of my brain that takes over in grizzly country. Jim was right. This time at least, arming myself against keystone predators was unnecessary.

Montana Highway 236 crosses the Missouri River at Judith Landing, an old steamboat port named after nearby Judith River. William Clark, a prolific renamer of Native American landmarks, scribbled Judith in his journals next to a hand drawing of the stream because he missed his fiancée, Julia Hancock, whom he called Judith. The Apsáalooke had called it *Buluhpa'ashe* (Plum River) for as long as anyone could remember. Today, Judith Landing has campsites and picnic areas, but no cell phone service. I arrived at the boat ramp around 2 p.m. and boiled hot water for tea while waiting for the shuttle back to Coal Banks Landing.

When a big canoe of sunburnt couples arrived, I waded out onto the ramp to help them land. They joined me under the tree, and one of the guys, Nathan, talked about how things have changed since the summer he worked as a ranch hand in the Bear Paw Mountains. "Celebrities have been buying up land around here," he said. "I've seen Edward Norton eating at the Grand Union Hotel in Fort Benton. Ted Turner owns lots of land."

Turner, the media tycoon, is one of the largest private landholders in America and owns ranches in Montana, including the Big Sky ranch, managed for bison and wildlife.

"Ted Nugent has a big ranch around here somewhere."

One of the women, tired and unable to mask her disinterest, looked at Nathan and said, "Yeah, Oprah has a lot of land up here too."

"Really?" he asked.

She smirked and popped the top on a grapefruit soda.

As ancient and static as the prairies of Upper Missouri Breaks National Monument seem, like the river itself, they are in constant flux, and like all North American grasslands, face a litany of dangers. I grew up in eastern Kansas, in a suburb where my sister and I used to walk to a small pasture of wild strawberries, black-eyed Susans, and wildflowers. Years later, I realized the micro-sliver of meadow on the steepest slope in our little world was almost certainly a remnant of never-plowed tallgrass prairie, a corridor that connected the suburban era of Ronald Reagan and New Wave music to ten thousand years of living natural history, when buffalo trod the same ground we sauntered on during the lazy, numbered mornings of childhood.

Everybody retreats to a fantasy world sometimes. In mine, it will never be too late to stitch together the tattered fragments of our native grasslands, and someday in that world a critical mass of people will realize the natural heritage of middle America and the Great Plains should be a living one, not relegated to history museums or Wikipedia pages. Today, the small pasture of wild strawberries and black-eyed Susans from my boyhood has been paved over with a private road. Rather than a seed that could become part of a prairied future, it was the last dying ember of tallgrass prairie in our neighborhood.

Upper Missouri Breaks National Monument is more than a seed. Whether or not it one day anchors a 3.2-million-acre biosphere preserve with buffalo that never encounter a fence, the breaks face more immediate threats. Noted river geologist Ellen Wohl uses the term "connected river" to describe rivers that are "fully connected to their surrounding landscape and richly diverse in form and function." She has written that, "in a connected river, pools and riffles alternate downstream, and riverside vegetation forms a mosaic of different types and ages. During floods, water flows across the floodplain, unimpeded by man-made structures."[25]

Connected rivers are wild rivers. In Upper Missouri Breaks National Monument, the Missouri is as "connected" as any twenty-first century grassland river. It's not without flaws. Native species like wolves and grizzlies, are gone, at least for now. Plains cottonwood (*Populus deltoides*), the keystone riparian tree of western rivers, has been declining since the Civil War when the soft wood was sold to fuel riverboat boilers. In 2005, the US Geological Survey and the BLM produced a study of cottonwood ecology to better understand the decline in the Upper Missouri Breaks.[26] At the time, most of the trees in the wild and scenic section were at least fifty years old. Cottonwoods can live for 150 years.

In early summer, the trees parachute millions of seeds into the wind and water. In freshly deposited alluvium away from other stands of trees, seeds germinate quickly. For the first few years, they need protection from floods, ice scours, and grazing livestock. Rising and falling river levels are required to produce fresh deposits of alluvium. In pre-European settlement times, these deposits were provided by snowmelt, spring breakup of ice jams, and thunderstorms.

From Lewis and Clark's journals, we know the Missouri used to be wider, shallower, and prone to more frequent changes in flow. In 1890, the first of five hydropower dams was constructed along the "Great Falls" of the Missouri above the breaks. Two water storage reservoirs were added in upstream tributaries in the 1950s. These structures have changed the river by stabilizing flows, limiting channel migration, and reducing the frequency of high-water events. Despite BLM fences around areas with seedling beds and water buckets placed so paddlers can water young trees, cattle grazing, along with winter ice scour, reduces the density and rate of seedling survival.

Centuries of change along the river have compromised its ecological integrity. But, because it is surrounded by world-class grasslands, the upper Missouri has suffered fewer degradations than its riverine brethren. That's why I went there first, to know how to recognize an extraordinary stretch of grassland river when I saw one.

In 2001, when President Bill Clinton designated Upper Missouri Breaks National Monument, some Montana landowners and ranchers believed he went too far with his presidential decree, that protection as a wild and scenic river was enough. President Donald Trump seemed to agree in 2017 when he ordered a review of all national monuments declared since 1996. Nothing changed because of the review. Today, Upper Missouri Breaks National Monument is still a place where people come to soak in the thick silence of the past and paddle the great river. But producers also use the land for cattle grazing, mineral exploration, and uplands farming.

This tolerance for mixed use is especially pertinent to grassland conservation farther east in states with less public land. In Kansas and Nebraska, for example, 98 percent of all land is privately owned, but the two states contain most of the world's remaining never-plowed tallgrass prairie, most of it on private property. The future of wild prairies in the Midwest and Great Plains will involve conservation on private lands, period. That's why I think the "buffalo entrepreneurs" at American Prairie who leverage strategies previously reserved for high-tech superstars and their shareholders could hold one piece of the puzzle in efforts to preserve and restore America's grasslands.

One of the Boy Scouts' chaperones thanked me for ceding the log shelter the night before. She lived in Denver and asked if I was from Kansas—she had noticed my "John Brown Ale" T-shirt, a nod to the wild-eyed abolitionist who got his start as the leader of anti-slavery guerrillas during the Bleeding Kansas period that led up to the Civil War.

She asked, "Did you ever go to the Outhouse?"

The Outhouse was an infamous punk rock club in Lawrence. The cinder block building in the middle of a cornfield hosted shows by Green Day, Iced-T, Nirvana, and other bands before they got famous. She said, "We used to go there when I was in college in Kansas City. I made out with the bass player of Trip Shakespeare in their van one night."

I remembered the band and might have been at the show.

A guy standing on a canoe trailer waved her toward him. "That's my husband," she said.

As she walked away, I noticed a small tattoo of the state flower of Kansas, a sunflower, on her shoulder blade.

"Is he a bass player?" I asked.

"Drummer," she said.

I thought of the Outhouse as the shuttle van bobbed up and down on the bumpy drive between Judith Landing and Coal Banks. My memories were a bleak Gen-X nostalgia at best: shaved-head punks drinking forty-ounce bottles of Colt-45 around a trash fire, kids stumbling into the cornfield to puke, the confusing aromas of hormonal underarms and loamy plowed dirt. But its surroundings are tattooed in my memories. The punk club, gone silent for decades now, sat in a wide delta near the confluence of the Wakarusa and Kansas Rivers. These would become my home rivers. Their confluence is steeped in mystery, a riverine treasure trove of natural and cultural intrigue. Now that I'd seen the breaks and floated the mother river, I wanted to return to Kansas with fresh eyes because things were happening there that symbolized river resurgence like on no other grassland river.

But I had one stop to make first. Spoiled by paddling among the unbroken prairies, I wanted to visit the "prairie inverse" of the upper Missouri: a river that traverses some of the deepest prairie soils in North America. Today those grasslands have nearly vanished. Instead of 90 percent, they were at best 1 percent unplowed, two extremes of grassland ephemera. The river seeps with unheralded potential and a hundred miles of sandbars few people ever set foot on. I'd noticed it shimmering like a golden serpent from the window seat of an airliner thirty thousand feet in the sky. Now, from the seat of a canoe, I would find out how deeply that river epitomized the paradoxes of twenty-first-century grassland rivers.

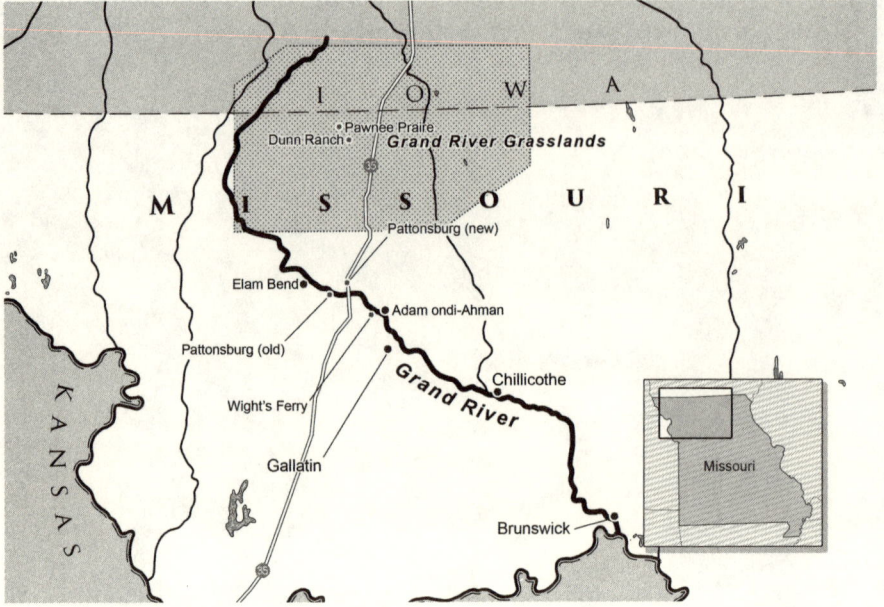

MAP 3 The Grand River

CHAPTER 3

GRAND

Buried in the middle of corn country, far from any ocean, the state of Missouri has a deep and intimate riverine history. The Missouri River itself is a blockbuster, as is the Mississippi on the state's sunrise border. When measured from its farthest source, the Mississippi runs almost thirty-eight hundred miles, making it the fourth-longest river in the world behind only the Yangtze, the Nile, and the Amazon, a river that is so dynamic and unexplored that even in the era of satellite imagery, debate remains over the longer of the two (traditionalists, however, favor the Amazon over the Nile).

Missouri has literary rivers. In the late nineteenth century, with the publication of *The Adventures of Huckleberry Finn*, *Tom Sawyer*, and *Life on the Mississippi*, the Missouri portion of the Mississippi River propelled not only Mark Twain but American literature more broadly onto the world stage. With it came the abiding mystique of America's big rivers and the country that surrounded them. In her 1962 novel *The Moonflower Vine*, Jetta Carleton described the unpretentious western prairies near the Missouri as a "region cut by creeks, where high pastures rise out of the wooded valleys to catch the sunlight and fall away over limestone bluffs." It's "a pretty country," she wrote: "It does not demand your admiration, as some regions do, but seems glad for it all the same."[1]

Missouri has P-A-R-T-Y rivers. Sparkling and spring-fed, woodland streams like the Current, Jacks Fork, Black, Elk, Meramec, Niangua, and the Gasconade Harbor nonnative rainbow trout and penetrate the forested moonshine hollows of the Ozark plateau, the crumbling afterglow of an ancient mountain range. During peak season they also host some of the biggest floating parties in the Midwest.

But Missouri also has grassland rivers, whose valleys spread like waves across sparsely populated farm country in the old dominion of the tallgrass prairie. Meltwater darlings fed from headwaters along the Continental Divide they are not. These low-gradient rivers drape the flat prairies and fens of southern Iowa, slow-grinding their way south until the Missouri stirs them into the silt-bearing slurry of mountain waters bound for the Mississippi. The Chariton, the Thompson, the Platte, the Nodaway, and the Nishnabotna dissect an agricultural plateau where the native prairies of pre-European settlement times are only a memory. But the Grand River stands out as the wild child of Missouri's tallgrass prairie streams.

Some of the rarest prairie remnants on earth flourish in its valley, grasslands connected to their beginnings in the glacial outwash of a lost time. During the early Pleistocene, ice sheets advanced and retreated twice across northern Missouri. The Grand River weathered its way through blankets of deep alluvium topped with loess, a rich soil ground to poultice long ago by those glaciers. The remaining soil is sometimes stacked in muscular cleavages sixty, seventy, or even eighty inches deep.

Unlike the vast prairies surrounding Upper Missouri River Breaks National Monument, or the shallow-soiled Flint Hills of Kansas that were spared because their cherty loams cracked steel plows, the deep-soil grasslands of northern Missouri were among America's "true prairies"—lush treeless jungles of sky that favored meaty grasses and forbs such as big bluestem, Indian grass, and rattlesnake master. These were prairies of such deep-rooted fecundity that in rainy years riders had to stand on their saddles to see ahead. This abundance created fertile soils perfect for farming and grazing cattle. Eventually, farming won out, and virtually all the deep-soil prairies were plowed.

"Prairie" is sometimes cast like a mist net to capture every possible variety of grassland. Ecologists, however, often reserve the term for tallgrass or *tall* prairie, a grassland subbiome hedged between the Great Plains and mixed-grass prairies that run from Saskatchewan south to Texas, and the eastern forests. The Dakotas east to Ohio and south to Texas are all tallgrass prairie states, some more so than others. In the early twenty-first century, never-plowed tall prairie is as rare as mastodon ivory. Scored against pre-European settlement figures, less than 2 percent remains, and that's a generous appraisal. Today, most unplowed tall prairies are lawn-sized relicts. Because no one remembers the sheer immensity of the millions of square miles of treeless, windblown biomass, our personal experience of it fades with each passing decade.

Near the headwaters of the Grand, however, cattle culture managed to hold on long enough to spare the unplowed pastures of the Nature Conservancy's Dunn Ranch Prairie, the single largest virgin deep-soil prairie left on the planet. The region has been well known for its prairies since at least 1755, when English botanist John Mitchell described it as "Extensive Meadows Full of Buffaloes."[2] In 2024, Dunn Ranch was one of the last remaining sites in Missouri where greater prairie-chickens—iconic symbols of the tallgrass prairie— still thrived, as did a tiny, bejeweled fish of unspoiled grassland waterways, recently reestablished in headwater streams of the ranch, a fish that symbolizes the umbilical connection between grasslands and grassland rivers.

The Grand River begins in southern Iowa as three separate forks—the West Grand, East Grand, and Middle Grand—that join near Albany in Missouri. From there, it courses 226 miles in a south-by-southeast heading through some of the least populated counties of Missouri before making its rendezvous with the Missouri River near Brunswick.

After paddling the upper Missouri in Montana, I wanted to apply my fresh knowledge and impressions to the Grand, the upper Missouri's "prairie inverse." Only about 1 percent of the Grand River's grasslands remain. But that's not *zero* percent. Hope remains. Its deep-soil prairies are no less important to America's wilderness legacy than the prairies of north-central Montana. Today, stakeholders such as the Nature Conservancy, the Missouri Department of Conservation (MDC), and other groups are crafting a broad vision to save the Grand's core prairies and restore tens of thousands of adjacent acres in a unique public-private partnership that could foreshadow future prairie rewilding across the Midwest. New science is critical to the vision.

Like its grasslands, long stretches of the Grand River itself are debased. Beginning in the nineteenth century, land speculators and farmers channelized its bends, felled its riparian forests, and polluted its waters with agricultural chemicals. However, unlike most rivers of the tall prairie, the Grand is free flowing; no dam has ever been built along its main stem. The story of how communities along the river fought flood control projects—in one case through a tremendous sacrifice—is testimony to the complicated relationship between the people of northwest Missouri and the river.

Both the upper Missouri and the Grand are part of the history of western expansion. While the Montana stretch of the Missouri was thoroughfare to Lewis and Clark, Maximilian, prince of Wied-Neuwied, and Karl Bodmer, the Grand River ran beside Adam-ondi-Ahman, a mysterious grassland bluff near Gallatin that played a central role in what historians now call the 1838 Mormon War, one of the darkest and most unsettling periods of Missouri history, a history that most Americans and many Missourians are unaware of today.

The Grand seemed like a buried gem: an unheralded, historic, and still relatively wild stream that embodied all the paradoxes of modern grassland rivers. To immerse myself in those paradoxes, I decided to start by visiting the deep-soil prairies of its upper basin to learn what rewilding efforts and the science of prairie restoration could mean for the future of grassland ecosystems in contemporary America. From there I planned to paddle one hundred miles of the river's most undisturbed sections. What my planning didn't prepare me for was just how wild the Grand River would be.

April in northwestern Missouri is a tempestuous month. One day the spring woods can be dressed up in blooming plum thickets, sweet williams, and redbud, and the next day frigid beneath a fleece of late snow. I was driving toward Dunn Ranch Prairie, a few miles south of the Iowa border, to watch the ancient courtship ritual of one of Missouri's last bands of greater prairie-chickens. The temperature was forecast to drop from the mid-sixties to below freezing overnight, so I mentally reviewed the hastily chosen clothing stuffed in my backpack.

It was already getting late as I raced twilight on the interstate. North of Holt—where it once rained a world's record twelve inches in forty-two minutes[3]—the deltas of two small creeks practically quivered with morels ready to pop like corks from the streamside humus. In the highway margins, spindly stands of common milkweed had begun to sprout, a delicacy the Ioway and Osage ate like asparagus (to remove the milky sap and mild toxins, *Asclepias syriaca* requires several long soaks, with a change of water each time, prior to the final boil; serve with butter and salt).

The roller coaster hills of glaciated Missouri could have been models for a Thomas Hart Benton painting. Vultures plied buoyant thermals, not yet situated after their recent migration from wintering roosts in Texas and Louisiana. Crossing from the final exurbs of Kansas City at Clear Creek into the *real* Missouri, a hill knocked the radio station from NPR to the Bott Christian radio network. Until the road dipped into a slight valley and columns of white smoke appeared low against the northern horizon, I hadn't seen a single prairie pasture during the entire drive.

Two hundred years before, this landscape would have looked very different. Early sources report that one-third of future Missouri was covered in tallgrass prairie. Because so much of it was lost, it is difficult to visualize today. But even though most people associate Missouri wild lands with the oak-hickory forests of the Ozarks—Mark Twain National Forest, Bell Mountain and Irish Wilderness, the Ozark National Trail, moonshine, and black bears— much of western Missouri and the lands north of the Missouri River were prairied in pre-European settlement times.

Walter Schroeder thought he could visualize that prairie, but he wanted to go one step further. Schroeder wanted to create an actual map of nineteenth-century Missouri prairie. What he needed was a time machine.

Schroeder grew up near the edge of the Ozarks in Jefferson City with his parents and grandparents, who had immigrated from Germany. When he was a young boy in the early 1940s, his mother volunteered for a group called Nature Knights that taught children about natural history. On one field trip, a professional conservationist took Schroeder and the other children to a stream and had them make field notes, recording the names of the species they found. This ignited his imagination. He went on to study geography and human ecology at the University of Paris and the University of Chicago before returning home to become a professor of geography at the University of Missouri in Columbia.

Retired now, Schroeder is in his mid-eighties but, with a full shock of white hair, looks much younger.[4] He told me that back in the 1970s, he was studying the cultural geography of Howard County, one of several counties north of the Missouri River that comprise "Little Dixie." Howard County was originally home to the Ioway and Osage peoples. In the seventeenth and eighteenth centuries, French *courriers des bois* ("forest runners") used the river as a thoroughfare to explore smaller streams and trap beaver. After the Louisiana Purchase, American immigrants came. Missouri would become a slave state, and before the Civil War some settlers brought enslaved people and established tobacco and hemp plantations along the Missouri River.

According to early accounts Schroeder and his students read, most of Howard County was originally covered in prairie. Schroeder wanted to see if there was better evidence for this, and a colleague told him that soil and land surveys of Howard County were conducted prior to Missouri statehood in 1821. Schroeder and his students found records not only for Howard County but six hundred and fifty volumes of surveys that spanned every county in the state. The surveys contained fine-grained information about vegetative cover down to the quarter section level. Schroeder had found his time machine. With enough effort, he and his students could use the surveys to reconstruct the ecological landscape and create a map of the pre-European settlement prairies of Missouri.

Howard County soil surveys, Schroeder told me, were conducted not long after the War of 1812, followed by a land survey six years later. These were different sets of surveys, with different goals. But what perplexed him were the differing accounts of vegetative cover for the same parcel between surveys. "I noticed that what the soil surveyors mapped as prairie soils derived from a grassland cover were quite a bit different from what the land surveyors had noted. I couldn't figure it out, why they were so different in their geographical distribution."

An archaeologist pointed out a six-year gap between the time of the soil and the land survey, a time when the Native peoples of Missouri were driven out of the state by American settlers. Schroeder said, "During that time, trees rapidly overcame some of the narrow patches of prairie, so the prairies were mostly gone by the time the land surveyors came in 1818. The rather quick invasion of Missouri prairies by trees fascinated me. Of course, it's because the fires that the Native peoples used to enhance the prairies stopped. Trees were an original feature of these landscapes, but the removal of Native Americans intensified the success of woody vegetation. Fire is a critical part of the story."

Because of provisions of the Louisiana Purchase, before land could be legally given or sold to settlers streaming west, the government needed to show that Native American, Spanish, and French title to the lands had been extinguished. This took time to litigate. Land surveys provided spatial support for the litigation but also functioned as advertisements for prospective settlers. Surveyors walked literally every mile of future Missouri, cataloging the vegetative cover at the cusp of an agrarian revolution. Survey teams fanned out across the state, along with support staff, cooks, chain managers, woodsmen, and hunters.

The surveyors used different terms for "prairie," a point of confusion for Schroeder and his students. They had to consult biographical data on individual surveyors to learn what terms they were likely to use to distinguish native grasslands from other coverages. "The

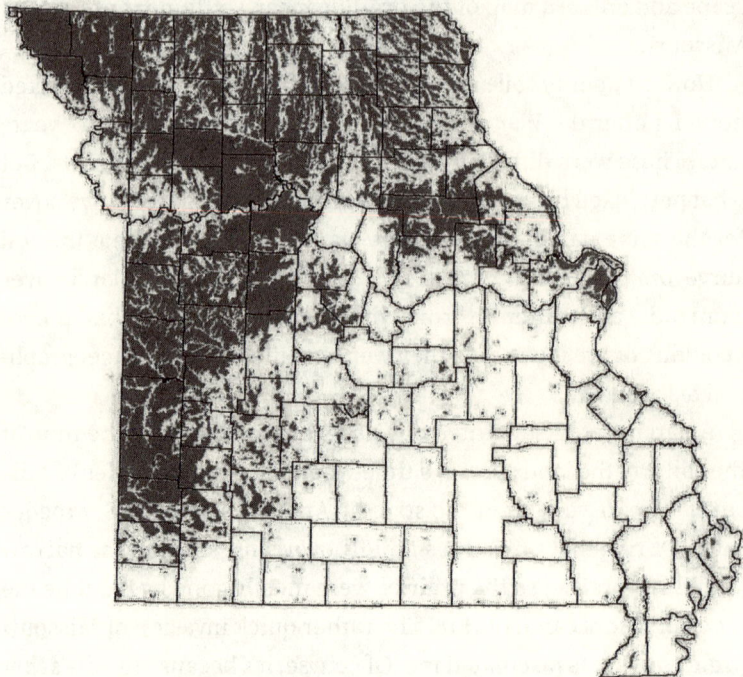

FIGURE 3.1 Presettlement prairie distribution in Missouri. Map courtesy of Walter Schroeder.

surveyor chose words depending, we found, on where he had done his education—even in 1815 these were educated guys," Schroeder said. "They came from Massachusetts, from Pennsylvania, from the Carolinas. They all had a different understanding of terminology. It's hard to account for this bias. They had different geographical traps that didn't adjoin each other."

A blue-blooded surveyor from Concord, Massachusetts, might use the term "barrens" for what a scrappier farmer's kid from Illinois would call a "prairie meadow." Schroeder took a conservative approach: "If they called it a grassland, I didn't call it a prairie, I used 'other.'" He said that if he were to redo the study, he would loosen his terse definition of what constituted prairie, although he says that not all grasslands in Missouri were prairies. "A lot of the open sections in the Ozarks were there because of Indian camps, and the French opened up land because they cleared the woods

around their lead furnaces [for making nails and bullets] for several miles and these were called prairies by the French. They were
grassy lands when the surveyors came through."

Using the surveys, Schroeder and his students mapped the entire state mile by mile. After tallying the data, they determined that
28 percent of pre-European settlement Missouri was covered in tallgrass prairie in the years between 1815 and 1855. He published his
"Map of Nineteenth-Century Missouri Prairies" in 1981.[5] The true
percentage was probably higher, closer to 35 percent of the state,
he said. Most of the prairies were in northern and western Missouri. Today, the parts of those prairies that have managed to survive are some of the rarest ecosystem fragments left in America.

I pulled off the interstate at Eagleville—population 277—the gateway to the prairie rangeland of northwest Missouri. Smoke drifted
like fog through trees leading down into the draws west of the highway. It was a better compass than the GPS in my smartphone, which
was getting no bars this far off the grid. Back in Kansas, thousands
of acres in the Flint Hills would be smoldering, in hundreds of separate prescribed burns that are a springtime ritual. The prairies
of northwest Missouri were connected to their distant brethren by
fire. Driving through the small village, a confederate flag flew beside a house near the town square. Welcome to Eagleville.

Not far outside town, the landscape changed. Grassy meadows
with cattle replaced plowed fields. On a hill steep enough to make
me think of the hiking poles back home in my closet, I turned onto
a gravel section road and switched off the ignition to take in the silence, accompanied by occasional meadowlark song, and the rush
of water in little rivulets from a morning rain. In an orange pasture
up the road, a few dozen shaggy cattle walked single file toward the
fading twilight. I grabbed my binoculars and saw they were bison,
not cattle. I'd found Dunn Ranch Prairie.

In a tight gravel driveway next to two empty bison trailers, I organized my mattress and sleeping bag in the backseat and strolled

out toward the road to look for the line of fire in the pastures. It was already cold. The damp earthy smell of the prairie made me glad for the winter bag in the car. Near the horizon at last light, a stationary silhouette began to move, and there was a flicker of light and a man with a cigarette walking up the road toward me. I was a little nervous, because I hadn't thought to ask permission to overnight in the parking lot, although there were no signs forbidding it. During the night I dreamed that the man banged on my window and invited me to look for prairie-chickens with him.

My travel alarm—the intruding headlights of a gold Volvo that rolled up next to my car—showed up twenty minutes before my planned wake-up time. I peeked out through the breathing hole in my bag. The driver, a woman with curly gray hair wearing a heavy-duty yellow Patagonia alpine jacket and a green wool scarf, flipped on the reading light and studied a map while her husband extricated himself from the Volvo, put his hands on his hips, and leaned way back, yoga-style. After about a minute of this, he walked across the driveway and let the door to the portable bathroom slam behind him. In the next twenty minutes, the parking lot filled with birders from the Kansas City suburbs eager to witness and photograph some of the rarest greater prairie-chickens in America.

A volunteer from the Nature Conservancy named Nancy opened the buffalo gate and our headlamps blazed six thin trails through the darkness. We hiked the half mile to the trailer blind in silence. It was important to get inside before predawn, so we didn't disturb the prairie-chickens, slumbering, I presumed, somewhere out in their grassy nests. Besides me, there were four couples and Nancy. Inside the trailer everyone hunkered down and unpacked tripods, light meters, camera bodies, and lenses by the armful from heavy-duty camera backpacks. I consoled my modest but loyal Canon. Nancy went over the rules one last time, although we'd all signed online waivers: no flash photography (external flashes bulged from the camera bags), no bathroom breaks, no lenses or body parts outside the viewing windows. No prairie-chicken calls. "Has that been a problem in the past?" I asked.

Any disturbance to this rare population would be troubling, of course. The greater prairie-chicken (*Tympanuchus cupido*) is the quintessential tallgrass prairie bird. The decline of the prairie-chicken parallels the decline of the tall prairie. Prior to European settlement, they ranged across 85 percent of Missouri, ranking second only to Illinois (100 percent) in total distribution across a state. In the twenty-first century, greater prairie-chicken numbers are declining throughout their range. Populations big enough to allow a hunting season persist only in Nebraska, Oklahoma, Texas, and, especially, Kansas. In Missouri, they've dwindled to near ghost status.

Prairie-chickens engage in a wild mating ritual each spring on flat patios of prairie called "booming grounds" or "leks." Male prairie-chickens gather to perform for would-be mates. Some of the leks are reused year after year for decades. The strutting males stomp their feet in rapid succession like a toddler having a fit. They wind around like a one-chicken conga line, entranced by their own footsteps. Then they stop, lower their heads, raise their tails, and, in one smooth gesture that provides hope for nature in the modern Midwest, inflate the orange interclavicular air sacs of their necks to produce a lonely, ancient moan that belongs, with the bugle of bull elk and the howl of the buffalo wolf, to the fragile and beautiful wonders that humans have done their best to destroy.

As late as the 1940s, greater prairie-chickens were common in Missouri. I once mentioned them to my grandmother, who lived in the small town of Versailles in west-central Missouri during the Depression. She told me she had seen them many times. My grandmother was a true Missouri girl, but I did not believe her. She'd never watched a nature show in her life. Since prairie-chickens didn't make the pages of *National Enquirer,* I was sure she had no clue what they were. When I said to her, "Prove it," she lowered her head, marched a few steps, and puffed her cheeks up! She had on enough rouge to even pull off an approximation of the neck sacs. During her girlhood, my grandmother and her sister had watched prairie-chickens booming in the pasture while walking to school.

By the beginning of the twenty-first century, however, Missouri prairie-chicken populations had plummeted into the hundreds. By 2024, the only two publicly accessible sites that still had living prairie-chickens were the Wah'Kon-Tah prairie near El Dorado Springs and Dunn Ranch Prairie. The birds' precipitous decline wasn't surprising. Missouri has many small relict prairie tracts, but prairie-chickens need more than cemetery plots and quarter section pastures to survive. They require large, high-quality grasslands.

Prairie-chickens were probably never extirpated at Dunn Ranch, but by 2011 fewer than two dozen birds remained. Since then, wildlife biologists have been trapping prairie-chickens on private sites in Nebraska and reintroducing them at the prairie. Even with that, the numbers aren't stable. With good moisture for two straight summers, by 2017 the resident population was up to three hundred. However, following five years of drought, in 2024 numbers had dropped below one hundred birds.

I'd started to nod off in the blind, but startled awake to the sound of rapid-fire photography. The first male prairie-chicken had arrived on cue as the sky began to lighten. Soon eight males were working the floor with forty hens huddled in groups of five or six. The hens feigned indifference—or were indifferent. A few snuck off into the stubble of last year's beans.

For the next two hours we watched an ancient comedy show, the males full of themselves and serious. At one point, a northern harrier swooped in for a closer look as a meadowlark pierced the morning with her shrill warning. In the blind, the coupled birders occasionally murmured requests of one another for more coffee or a different lens but otherwise were silent and intent on prairie-chicken photography. Nancy told us that a harrier once managed to lift a prairie-chicken to the height of a grown man before it broke tackle and fell back into the huddle with the hens. Prairie-chicken cliques spread out over the field for about a quarter mile. There might have been forty birds in all.

FIGURE 3.2 Greater prairie-chicken at Dunn Ranch Prairie in northwest Missouri. Photo by the author.

After two hours, the hens flew off the lek. At first the males didn't seem to notice and kept jiving for a minute or so like wind-up toys, but they eventually got the message and flew west with the others. The show was over.

The reason greater prairie-chickens still have hope in Missouri is because Dunn Ranch, already the largest remaining tall prairie in northern Missouri, is part of a larger vision: the 70,000-acre Grand River Grasslands Conservation Opportunity Area (or just Grand River Grasslands). The ranch, with its thousand-acre core of never-plowed prairie, is the anchor tenant of the project. The MDC emphasizes prairie restoration in the region. If private landowners want to replant prairies, they can get increased cost share from MDC and access to experts to help. The Grand River Grassland is an unusual experiment and the kind of partnership that, if it succeeds, might one day become more common across the prairies and plains. It hinges on a mix of private and public land ownership and the latest in cutting-edge environmental science and technology.

MDC natural history biologist Steven Buback is one of the experts. He told me the project is "a way to focus the resources of MDC in a single area to create a large grassland to support species that require big expanses."[6]

Though 70,000 acres might seem small compared to the proposed 3.2 million acres of the American Prairie Reserve, Buback emphasized how rare such a large grassland in the deep-soil regions would be. "Dunn Ranch Prairie is on the glaciated till plains, an ecoregion that extends from northern Missouri to parts of Illinois and Wisconsin. Essentially, this is what we think of as the corn belt. The tallgrass prairie in the glaciated plains is on much richer soil, they are very deep productive soils. This was where the prairie was at its most productive."

Besides acquiring the remaining unplowed prairie, Buback said the Nature Conservancy has focused on keystone species as one cog in its preservation and restoration efforts. "Prairie-chickens are a good poster child, because when you save prairie-chickens, you save habitat for a lot of these other species, some that we don't even know are there—they may be rare or common, but they've never been surveyed for or documented."

Most of the land in the target area is privately owned. Local ranchers understand the value of native grasslands, and most have been supportive after they learn about the program. At first some cattle farmers considered the ranch a "weed patch," Buback said, and some thought the federal government owned it (the Nature Conservancy is a nongovernmental organization; the ranch is privately owned). Unlike the American Prairie Reserve, none of the ranch's neighbors have been openly antagonistic to the project. Buback said, "The reason the tallgrass prairie doesn't get talked about as much is because most of it is gone. We've lost 99 percent. It's not a defining feature of the landscape and it hasn't been for a hundred years. What's imprinted on people today are prairie restorations and little pockets that we have remaining. But these pastures are beneficial for nature and commercial cattle operations if managed correctly."

Most of the restorations have surrounded the virgin core at Dunn Ranch. Buback said, "Everything that we've done has been an approximation of a restoration. Even the oldest restorations don't have the suite of species you might find on a native prairie. In the past we've focused heavily on plant material. So, if we're lucky and we can afford it, we'll get seeds to a couple hundred different species of plants, throw them out onto this piece of property, and then manage it."

Buback said that restoring plant diversity has gone well, but diversity isn't the only measure of success. "At this point in history, a restored prairie is still noticeably different from an intact prairie, especially to a trained eye. Just in terms of plant composition, in terms of the species that are dominant. In a restoration, plants like big bluestem, Indian grass, and tall goldenrod thrive. There are other species, like the eastern prairie fringed orchid, that don't take well to restoration because they are dependent on things other than just the vegetation characteristics, like mycorrhizal associates in the soil, bacteria, invertebrate communities, which traditionally haven't been a focus of restoration."

According to Keith Bennett, who specializes in native seed harvesting and prairie restoration for the Nature Conservancy, the organization started converting fields of fescue at Dunn Ranch back to prairie in 1999, reseeding and using prescribed burns to manage invasive plants like eastern red cedar and honeysuckle.[7] A few years later, they began to record species they hadn't expected: delicate tallgrass prairie forbs such as sensitive briar and downy blue gentian whose seeds and roots had been slumbering beneath the soil for decades. A bison herd established in 2011 has grown to 150 animals. Unlike cattle, the buffalo only eat grasses, according to Bennett. They leave forbs and small shrubs that grassland birds use for cover and food.

Bennett said they work with Drake University ecology and botany professor Thomas Rosburg on vegetative surveys that monitor the rewilding efforts. One technique involves marking off ten-by-thirty-meter plots and counting the number of species found in each plot. Their best-restored transect topped one hundred species, and as their techniques improve, they'll be able to measure the success in test plots.

As part of the Grand River Grasslands work, the ranch recently began a grass bank, a program that allows area ranchers to use Dunn Ranch to graze cattle while the rancher's pastures are allowed to rejuvenate. Bennett said, "They can come in early spring and graze for sixty days, then come back in fall and graze another sixty. We don't charge rent, instead we require ranchers to implement conservation practices for prairies on their land. It could be planting warm season grasses or cutting trees. It depends on the pasture."

What worries Bennett the most is the noxious invasive, *Sericea lespedeza*, a cancer to restored and native tallgrass prairies across the Midwest. To prairie people, lespedeza is a swear word. Few fighting it in the field give it the courtesy of a common name, preferring the Latin. The perennial legume from Asia was introduced as a cover crop to control erosion a hundred years ago, but today it is depression incarnate to native grasslands. Fire alone doesn't kill it. Bennett said, "One bush will produce a hundred thousand seeds per year. That's a twenty-year seed base. If it goes to seed once, you've got to check the area for twenty years and not miss anything. It's a generational battle." Triclopyr with and without fluroxypyr, picloram with fluroxypyr, metsulfuron, glyphosate, imazapic applied at various stages of the plant's life cycle using backpack sprayers by riders on ATVs or horseback have all been tried. None has completely eradicated the plant from large areas, and chemicals like triclopyr, though they have a short half-life, must be applied with great care, especially around water. New protocols are under constant development and Bennett is hopeful that restorationists can stay ahead of the alien legume until more effective protocols become available.

Despite these measures to control *Sericea lespedeza*, the challenge highlights the ongoing struggle to protect unplowed prairies and prairie restorations. The effectiveness of these efforts can often be gauged by the health of indicator species, such as the greater prairie-chicken, whose presence at Dunn Ranch signifies the high quality of the prairie habitat there. As indicator species go, greater prairie-chickens and American bison are both showy exemplars— they help people understand what prairie *means*. Other indicator species are less conspicuous. A prime example is *Notropis topeka*, the Topeka shiner, a tiny fish of tallgrass prairie waters, whose presence divines proximity to pristine grasslands no less than the greater prairie-chicken. The little fish epitomizes the unbreakable bond between grasslands and grassland rivers.

The Topeka shiner is a minnow that evolved in the tallgrass prairie streams west of the Mississippi River. They are much pickier in their habitat requirements than the other small fish they commonly school with, such as the central stoneroller, redfin, sand shiner, common shiner, and red shiner. Topeka shiners are typically found in cool headwater streams with high water quality. They have a snazzy black chevron at the base of their tail fin and are prone to scoliosis. They spawn in silt-free waters over the nests of sunfish, particularly those of the orangespotted sunfish, a member of the Lepomis family that sports a striking orange and turquoise mohawk of dorsal fins during the spawning season. The original range of the Topeka shiner was restricted to upland headwaters of tallgrass prairie rivers, where springs and groundwater seeps cooled shallow flows that crossed clean gravel and sand. Kansas and Missouri—including the entire Grand River drainage—once provided prime habitat, as did parts of Nebraska, South Dakota, and Minnesota. The near eradication of the tallgrass prairie has driven the Topeka shiner to the brink of extinction.

Just as there are various types of grasslands, there are also different kinds of grassland rivers. According to Christopher Janssen, a life scientist at the Environmental Protection Agency, the quality of Topeka shiner habitat directly reflects the quality of the surrounding prairies. "Most tallgrass prairie rivers don't originate in the mountains and are different than shortgrass rivers farther west because of the topography of their floodplains. They typically don't have bedrock holding the stream in place, so there is more room for these rivers to move through their floodplain where the channels can migrate."[8]

According to Janssen, the story of tallgrass prairie rivers begins with their low gradients: "Many prairie streams don't follow the traditional run, riffle, pool structure of other rivers. There's just a consistent glide for miles and miles, because of the low-gradient profile." Tallgrass rivers flow through wide, flat floodplains with bottoms that gently slope into the uplands. They often change course, creating oxbow lakes and sloughs off the main channel. The main stems of these rivers are muddy due to the natural erosion of loess and clay soils, while the spring-fed upper regions support clear-running tributaries.

Some tallgrass prairie fish, such as the johnny darter, creek chub, and fathead minnow, are generalists that can tolerate some environmental degradation. Janssen said, "the Topeka shiner, though, is really selective. Their downfall has been the disruption of bedding and nesting sites, which can't be in turbid waters. The level of infiltration in a native prairie is just crazy," he said, referring to how prairies retain water during rainfall events. "Converting prairies to agriculture results in sedimentation of streams because of increased velocity of runoff from tilled fields. It changes the hydrology of the system. This increased turbidity, and perhaps higher levels of toxins and changing pH, has affected aquatic insects and bottom-swimming minnows like the Topeka shiner."

In the early 1990s, fish surveys revealed that Topeka shiner populations had dwindled to less than 10 percent of their original historic range. They disappeared from areas where prairies were plowed, spring runs were cut off from downstream rivers, and headwater streams became silted in due to cultivation, urbanization, and highway construction. Endangered species status was granted in 1998. By the turn of the century, the greater prairie-chicken had emerged as the avian symbol of the tallgrass prairie, while the Topeka shiner became the ichthyological mascot of tallgrass prairie rivers.

In 2017, the Nature Conservancy acquired the 217-acre Little Creek Farm near Dunn Ranch, which includes a portion of the Little Creek headwater. Alongside three other spots in the upper reaches of the Grand River watershed, the farm was selected as a reintroduction site for the Topeka shiner in an effort to revive the species where it had disappeared. Eroded stream banks were angle-sliced to restore the original gentle slope, and broken culverts were removed. This action reconnected creeks to cold springs in upland pastures, reopening historic spawning routes that had been previously blocked for the fish.

Conserving headwaters also brings downstream benefits. The Grand River is the largest phosphorus contributor to the Mississippi River basin. A program associated with the reintroduction effort collaborates with local farms to promote fertilizer application practices that specify the optimal rate, timing, and location to minimize nutrient runoff into the watershed. This initiative not only supports the Topeka shiner and a diverse range of fish species but also enhances the health of the main stem of the Grand River, located five miles downstream. This underscores the significance of leveraging the heft of the Endangered Species Act. As Steven Buback told me, "If you want to effect change on a big prairie river, start at the top. That's the most important thing to remember."

A week before Halloween, my friend Ted helped me shuttle my canoe between Gallatin and Albany for a solo float on the upper Grand, where the water runs a steady three feet deep in times of good water. This stretch includes the unchannelized Elam Bend Conservation Area.

Northwest Missouri's backroads often run parallel to larger state and federal highways. This network of "letter" or "double letter" roads (like Highway EZ) winds through a grid of gravel roads laid along section lines, crisscrossed by an even more rudimentary network of earthen trails gradually returning to the land. Seen from aerial photographs, these paths resemble scars from successive surgical procedures using increasingly advanced technology over decades.

The desiccated remains of four big catfish dangled from a sign atop the hill where we exited the interstate, welcoming us into the valley of the Grand River. The road descended through fields of soybeans tinted in rusty yellow hues. In a nearby field, an aging plywood board was spray-painted with the words, "Pattonsburg was here." By midday, we arrived at a clay cliff near Albany and, with the aid of ropes, carefully lowered my canoe and gear down the bank. There was no trail to follow, nor anyone to assist us, so we improvised, sliding swiftly down on our backsides. Ted, sinking into the soft clay up to the top of his shiny black Doc Martens, shoved me and the canoe off into the stream.

The river at Albany was navigable, a good sign given my late start, leaving only a few hours of daylight for paddling. With my best aluminum paddle, I dug into the current using the feathered blade, tracing J-strokes until the canoe was slicing through the thin channel. A cold front loomed, set to arrive soon.

The first route-altering root wad blocked the second bend shortly after the put-in. The Grand is one of the last significant undammed prairie rivers in the Midwest, but that doesn't mean it is unchanged from pre-European settlement times. Midland rivers were once straightened by slicing out their bends. That engineering practice had been going on in Europe for at least two thousand years when settlers reached the Grand. In the nineteenth and early twentieth century, laborers used horses, plows, and shovels to remove bends in large sections of the Grand and most other Midwestern rivers to drain wetlands, enhance navigability, improve grazing, and control flooding. Storm waters travel faster down a straight corridor, and ad hoc farm levee construction and channel alterations create tall stream banks that quickly crumble and erode. Studies show that straightening rivers can diminish fish biodiversity by eliminating riffles, pools, gravel and sand beds, and backwaters.[9] Riparian forests can collapse into the channel when banks crumble. That narrative unfolded over the next five miles.

Deadfalls clog the upper Grand: collapsed hedges of popping sandbar willow, maples, sapling sycamores, and the bodies of ancestral cottonwoods and burr oaks. Downed trees on a twisty river are dangerous. I stowed my ropes so that nothing bulged over the gunwales for loose branches to snag. On blind curves it is important to stay as far to the inside as possible, or otherwise hop out of the boat and investigate from land. I gunned the straights, building momentum using a sit-and-switch stroke pattern—three strokes on one side and then "hut!"—really digging in to stay in the main channel and then hopping out on the upstream end of sandbars to scope the curves on foot. I had to ferry the boat and my provisions through tree jams three times by rope.

After five hours, the logjams subsided. I scraped through gravel riffles near Elam Bend, a beautiful unchannelized section of the river. The tracks of river otters, beavers, great blue herons, bobcats, and muskrats annotated a subdivision-sized spread of sand landscaped with a nursery of sapling cottonwoods. The wind had begun to whip through the channel and grains of sand blew across the river and stung my face like airborne kidney stones. Two miles later I pulled on a jacket; the cold front had arrived with about an hour remaining before sundown.

Shopping for the perfect sandbar, I hoped to find one more beautiful beach spread like the ones at Elam Bend, but beyond the bridge at Old Havana Trail the sandbars disappeared along a monotone, straightened section of the river. After another mile I tried to turn around and muscle back upstream to the bridge and a roadside campground, but the wind cussing down the channel spun me back. A second attempt to turn nearly capsized me sideways, forcing me to pry the boat back into the current with effort.

There was nothing to do but paddle hard and hope for another pullout. A thin mustache of sand below a limestone face that stubbled the river bottom with gravel appeared after a mile. A wooded terrace above the cliff would have offered protection from a westerly, but the wind had shifted around to the north. The hills above the river provided no solace.

I selected a spot in smooth sand and scouted a little trail into the brush before making camp. It was miles from the nearest paved road. Stepped bluffs flanked the river valley to the west—the same bluffs, possibly, where Daniel Boone spent a winter in a small cave with Derry Coburn, an enslaved man, hiding from the Osage or Sac and Fox, who had warned Boone to stay away.

I'm not squeamish about sleeping outside. Luckily, no tent-crunching grizzly has ever come sniffing around one of my camps, but I once bivouacked in a soybean field and woke when a coyote stepped on my leg. We both yelped. But the wind was getting stronger, and I wondered how long I could last on the open sandbar. It took thirty minutes to set up my tent and tie everything down. I sat inside with my arms wrapped around my knees, motionless and tired. Soon I heard the canoe flip over, rolled by the wind. The rain fly flapped like Tibetan prayer flags. Still hoping to salvage the soft sand for a campsite, I dragged the boat around and hedged it behind the tent for a windbreak. It took all my remaining rope to tie it down. Five minutes later the canoe bucked hard in a strong gust like a spooked horse. Night was falling, but if the canoe flipped over it would crush not only the tent but me inside.

At twilight there was no choice but to break camp and move up onto the terrace in the woods, a move that took twelve trips along the sandy trail. I dragged the tent, poles in place. It was more like hauling laundry than portaging. The canoe stayed by the river, hammered in with three more stakes, and cinched down tight.

My cottonwood camp was no better. The incessant smacking of the rain fly made reading, much less sleeping, impossible. Dispirited, I spooned down some soggy lentils, and after thirty minutes trying to read, switched off my light and fell asleep.

A loud bang woke me. "My canoe!" I found the headlight, pulled on boots over my long underwear, and staggered down to the sandbar where the boat was staked in. The canoe was fine, but, not keen to take any chances, I dragged it up the sandy trail into the woods and past the tent, dropping it somewhere in the darkness. "Goodnight canoe," I whispered, "may we rest."

For the next hours, packed isobars of the arctic system descended on Elam Bend. At 10:45 p.m., after a silence that lasted long enough for me to hear a cow moaning across the river, I heard what sounded like jets, followed by the unmistakable pop of a tree trunk bursting from its torso and tumbling into the water. It had to be one of the great cottonwoods across the river. A minute later another trunk torqued and the earth below me trembled. I pulled my backpack over my body for protection as branches and leaves crashed down.

The chaos continued for what felt like twenty minutes, though it might have been only two. The storm peaked in one epic downburst with the sound of metallic shearing followed by a violent pounding in the woods. Suddenly, my tent was like a toadstool exploding under a kid's sneaker, with poles snapping and the tent deflating around me, flattened fortunately by wind and not wood. I laid there in the dark, collapsed mess and dug around in my pack for a protein bar. Taking a bite, I remembered a midnight storm in Kansas years before, when I had watched through the hail-deformed mesh of my friend's tent as a ragged funnel cloud, illuminated by sheet lightning, spun directly above us.

The wind eventually grew tired of tormenting Elam Bend, so I dug out some duct tape and splinted two of the busted poles, slowly reassembling my shelter.

In the morning, I hiked out to where I'd left my canoe—the *Bluestem Rises*—or at least to where I remembered leaving it. I checked all around the little clearing and in the surrounding groves, under the two huge cottonwoods that had blown into the river, along the shore below the bluff, and across the full length of the sandbar. I hiked up the trail and checked along the bank for half a mile.

A four-hour search turned up nothing. The canoe was gone.

The shearing and pounding noises probably meant the canoe had flown end over end out of the woods and into the river. Where was it now? In a hunter's camp? In the back of a pickup bound for Iowa? Sunk in pieces below the cut bank across the river? Folks at the MDC put word out to people who traveled the river, but to this day, the canoe remains MIA.

After the boatless search, I packed everything up and hiked out of the woods to wait beside Highway Z for my wife to make the long, short-notice drive from Kansas. Consulting my maps, I noted that below Elam Bend, the river entered Daviess County and passed by the grassland at Adam-ondi-Ahman. Sitting on the cold ground beneath a living wreath of wild Missouri bittersweet, I wondered what surprises the Grand River might have when I returned to learn about its role in the almost forgotten 1838 Missouri Mormon War, a brutal precursor to one of the central stories of western expansion.

After the ice thawed the following spring, I stood on a hillside northwest of Gallatin staring down at a green ribbon of forest where the Grand River was flowing through a lush valley sparkling with morning dew. Cows grazed behind a long, white-washed wooden fence. Red-eyed vireos called from an ancient burr oak. My friend Jay and I had stopped at the spot to organize our gear on a picnic table before setting out to paddle through Daviess County, the next section of river below Elam Bend.

Today, the neatly manicured park and surrounding grassland called Adam-ondi-Ahman is a shrine that commemorates the bluffs where Joseph Smith Jr., founder of the Church of Jesus Christ of Latter-day Saints (LDS), came with a surveying party in May of 1838 to organize a new city. To learn about the history of the LDS church in Missouri before my trip, I spoke with historian and Brigham Young University professor Alexander Baugh about the series of conflicts between LDS settlers and Missourians in 1838 and 1839 that escalated into armed confrontations along the banks of the Grand River.

Baugh, whose research has focused in part on the "Missouri period" of the LDS church, told me that the state and the Grand River valley are important in LDS History. He said, "the LDS church has peculiar beliefs about Book of Genesis geography" that aren't shared by all Christians. "Joseph Smith, Jr. taught that the Garden of Eden was located in modern Jackson County, Missouri, the county where Independence is today. After they were expelled from the Garden of Eden, Adam and Eve moved north to Adam-ondi-Ahman. The name means 'Adam with God,' and their posterity remained there until the Great Flood."[10]

Smith, the LDS prophet, said the church would someday build a Temple of Zion in Independence that Christ would visit in the Second Coming. Smith and his followers spent much of the 1830s in Kirtland, Ohio, but during that decade, LDS populations swelled in Jackson County, close to the future Zion.

Most non-LDS Missourians near Independence were fiercely proslavery and anti-Mormon, in part because of the nominal abolitionist stance held by most of the LDS (which might sound strange, since the church would later teach that African Americans were cursed with dark skin as the "mark of Ham" until they abandoned the policy in 1978). In July 1833, a mob of Missourians destroyed the press and offices of the church newspaper in an attempt to force church members to sell their landholdings and leave the area. That November, Missouri lieutenant governor Lilburn Boggs, a veteran of the War of 1812 from Kentucky who had moved to Missouri to marry Daniel Boone's granddaughter, met with Mormon leaders. As a gesture of goodwill, he asked them to persuade church members to surrender their arms to Boggs's militia. That same night, vigilantes rode across Jackson County, attacking defenseless Saints, pistol-whipping men and boys, assaulting women, torching haystacks and crops, and leaving more than a thousand people homeless at the onset of December. Mormons there died of exposure during the harsh winter of 1833.

News of the attack and the Mormons' winter of death evoked public sympathy. The Missouri state legislature took note. After bicameral debate and negotiation with church leaders, Missouri authorities persuaded the church to relocate forty miles to the north of Jackson County, to the new Caldwell County. Missouri had set aside the county as a "reservation" for Mormon settlement.

When Joseph Smith had moved from Ohio to Missouri, he traveled overland from Kirtland to St. Louis, then proceeded up the Missouri valley. From there, he rowed up the Grand River with a small company to Wight's ferry in Daviess County near Gallatin. LDS member Lyman Wight had recently settled there and established a ferry to transport people across the river during high water. Leaving their boats to hike up into the limestone bluffs east of the river, Smith discovered a tight semicircle of flat rocks. He would eventually show the spot to numerous followers, but on that day, he reportedly said, "This is the valley of God in which Adam blessed his children and upon this very altar Adam himself offered up sacrifices to Jehovah."[11] Smith spent June of 1838 surveying the area and laying out a new community. Almost overnight, Adam-ondi-Ahman became the center of LDS activity in Daviess County.

Our plan was to paddle to the unmarked Wight's ferry site, where the soft-bottomed river crosses over limestone bedrock. Jay and I launched the canoe from a put-in beneath the bridge south of

Jameson. Twenty days since the last rain the Grand was running about two feet deep. We waded into the stream from a spongy bank piled with sugary white sand on the western edge of Adam-ondi-Ahman. We managed to stick the boat four times before locating the only cobble bar north of the big easterly sweep—the site of Wight's ferry. There is no trace of the ferry today; it was only in operation for a few years. On this day, nobody would have needed a ferry, it would have been easy to walk across the river with horses and wagons. Jay said, "We'll be lucky to make two miles per hour in this current."

This was the first place on the river that provided a clear view of the farmed valley. We had launched from the apex of a finger loop that curled around a bottoms that once must have been laced with wetlands. In the Depression, planners considered building a "Pattonsburg Lake" at the bottom of the bend, but the project was abandoned. The lake would have submerged most of Adam-ondi-Ahman. Today, this section has an intact riparian forest and channels that weave around each other like multiple vortices of a tornado. Mussel shells litter the gravel bars. As we sat to eat our lunch on a sycamore log looking out onto the beautiful valley, it was hard to imagine the chaos and terror that occurred there in the late 1830s.

For a while, relationships between Missourians and the LDS at Adam-ondi-Ahman were peaceful. That changed in the late summer of 1838. On August 6, Gallatin held an election. When voters arrived to settle a close runoff that included an LDS candidate, Missouri settlers tried to intimidate the Mormons from voting. Trouble broke out. Sampson Avard, a member of the church's high council, had organized a secret paramilitary group, the Danites—a sort of Mormon Hells Angels—to retaliate against the brutality church members had experienced in Missouri. On election day, Avard's Danites got into a brawl at the Gallatin polling site, beating Missourians with oak clubs and wounding more than a dozen men. Over the next two days, church leaders attempted to quell the rebellion. Concerned that reinforcements might arrive to defend LDS settlers, a Missouri militia and a mob of civilians began to mass outside Gallatin.

Fearing a rout, the Danites and the church's official security force began clashing with the Missouri mob, and in the next weeks skirmishes broke out and homesteads on both sides were ransacked. Eventually, hostilities escalated, and now-governor Lilburn Boggs issued Missouri Executive Order 44, the infamous Mormon "Extermination Order." In a short letter to General John B. Clark, Boggs wrote, "The Mormons must be treated as enemies, and must be exterminated or driven from the state if necessary for the public peace."[12]

Two men were killed, one from each side, at the Battle of Crooked River in Ray County. The Missouri militia massacred the Mormon community at Haun's Mill, killing eighteen men and boys. On Halloween, the militia attacked the LDS settlement at Far West. Meanwhile, the militias drove Daviess County Mormons up into the bluffs above the Grand River.

To prevent a slaughter at Adam-ondi-Ahman, Smith agreed to parlay with militia leaders. He was arrested and sentenced to a summary military execution, but Brigadier General Alexander Doniphan refused to honor the orders, declaring it an illegal court-martial. Smith and others were jailed in Independence and eventually in a small prison at Liberty.

The militias forced all remaining Mormons to abandon their property and leave Missouri. Future Mormon prophet Brigham Young led the Saints back across the Mississippi River to the last LDS capitol east of the Rockies at Nauvoo, Illinois. Soon after, Adam-ondi-Ahman was abandoned.

Jay and I left Wight's ferry and started paddling toward Gallatin. Gradually the channel turned to rock and cobble, and we paddled through low voltage rapids, then took a sharp left turn into a straight easterly run and entered the most beautiful section of the river.

We'd been spooking bald eagles for a couple of miles, but finally managed to float by an eagle perching motionless in an elm tree on river right. Jay called out two pileated woodpeckers cackling across the river. We paddled hard to catch up, almost flipping the boat because we weren't in sync. "I'll paddle, you pileate," I said to Jay.

At the point where the river veers south away from Adam-ondi-Ahman and back toward Gallatin, a dozen tree-sized branches reached out across the river from a sycamore. The branches were full of birds. I said, "Heron rookery." We didn't breathe and stopped paddling as the massive white tree pulled us in like gravity. Without our field glasses, it was a shock when not herons but twenty bald eagles lifted in unison, two remaining in the tree to watch us float by. I once saw two hundred eagles gathered around a single spot of open water on a public reservoir in Kansas, but this was the most eagles I'd ever seen in a single tree unrelated to a migration.

In April, 1839, Joseph Smith Jr., escaped imprisonment in Liberty and left Missouri for good, settling with the rest of the LDS in Nauvoo. Five years later, he was killed by a mob who broke into the prison where he was then awaiting trial for treason. Following Smith's death, Brigham Young led the surviving members of the church to Utah along the Mormon Trail in 1846 and 1847.

Two years after his term as governor expired, a sniper shot Lilburn Boggs in the head as he read the newspaper in his Independence home. Mormons were dry-eyed. Boggs eventually recovered and later got rich selling dry goods during the California gold rush.

The Grand River Historical Society in Chillicothe has several exhibits dedicated to the 1838 Missouri Mormon War and the Gallatin Election Day Battle. When I visited, one of the volunteers took me aside. "We're really proud of our museum, but I've lived here all my life, and I'm not proud of what happened to the Mormons in Missouri." She's not alone. In 1976, Missouri governor Kit Bond officially apologized and nullified Boggs's extermination order.

The near-Olympian prairies of north-central Montana might seem an unfair comparison, but the prairies of the Grand River Grasslands are trembling with potential. To restore lost ground and revive a semblance of their pre-European settlement biodiversity, and for greater prairie-chickens and other keystone species to establish self-sustaining populations, rewilding will be necessary.

If only tallgrass prairie restoration was the mere masonry of sodding a suburban lawn. It's far more complicated. Grassland restoration in America began during the Dust Bowl, in the Great Plains, when Congress included funds in the Bankhead-Jones Farm Tenant Act of 1937 to purchase the thousands of Dust Bowl–stricken farms and ranches that eventually were stitched together to form the national grassland system. The lethal soil erosion that caused black blizzards made shortgrass prairie restoration a pragmatic affair, and the Soil Conservation Service sowed seeds in an ad hoc manner to reestablish vegetative cover in the 1940s and 1950s. Ecosystem-focused restoration ecology emerged around the same time, with the broader goal to assist or—in the case of tallgrass prairies—reestablish degraded, damaged, or destroyed grasslands. Hard science didn't factor into grassland restoration until the 1970s.

Defining "success" is one challenge. Restorationists must know what they're shooting for and have a methodology to measure how well the restored ecosystem matches the original one, ideally something like the Freese Scale used by American Prairie. They need to develop specific methods, planting strategies, and management techniques to achieve well-defined goals. Today, tallgrass prairie restoration is an active field of research with conferences, distinguished professors, grad students, and interns who spend summers cataloging species, running forb censuses, learning to adapt farm equipment to undo the effects of farming, and drinking beers together at sunset, worn out to the bone. But as scientific disciplines go, restoration ecology is still in its infancy.

Starting in the 1980s, research goals began to expand beyond simply reintroducing the most common prairie grasses to encompass rare forbs, invertebrates, small mammals, buffalo, and regular prescribed burns. But by the second decade of the twenty-first century, prairie restoration still suffered from a fundamental problem: no restored tallgrass prairie has passed ecological restoration's Turing test. It's easy to distinguish even a forty-year-old prairie restoration that's been seeded with hundreds of species from a native prairie that shares its border. The forty basic grasses and forbs do well enough, but the other sixty late-successional forbs and legumes and delicate grasses never make it.

Restorationists have now shifted their focus to the microscopic realm in a quest for a "missing link" that could one day enable the replanting of tallgrass prairie with the ease of planting lawns. Soil analysis is a critical component of this work. Measuring soil chemistry can determine what species and prairie types are suited for a given parcel. Some scientists focus on establishing adjunct invertebrate communities that match those found in native prairies.

Plant and fungal ecologist Liz Koziol believes that she knows another of the microscopic missing links. Beginning in the 1990s, prairie ecologists turned their attention to subsoil ecosystems. Fungi and mycorrhizal fungi are components of the prairie's microbiota, the hordes of beneficial microscopic wigglers that perform duties, similar to a person's microbiome, that scientists are only beginning to understand. The thin thread-like hyphae that compose the mycelium of mycorrhizal fungi are as delicate as spider silk. Plowing a prairie or field chops native fungi to death.

Koziol, alongside scientists at the University of Indiana, the Land Institute, and the Bever/Shultz Lab at the University of Kansas, has used endomycorrhizal fungi that she carefully raises from samples collected in unplowed prairies (Koziol emphasizes she doesn't damage the prairies to get the samples) to inoculate played-out agricultural soils prior to reseeding. She's found that her fungi cultivars have no effect on easy-going species like big bluestem or black-eyed Susan, but when sown together with finicky delicates like the federally threatened Mead's milkweed, the results are stunning. Late-successional species that almost never survive in replanted prairies without the support of mycorrhizal fungi are showing up again. When scientists learn to develop similar inocula inexpensively at scale, it could fundamentally change how tallgrass prairie restoration is done.

The 70,000 acres that surround Dunn Ranch Prairie are a reservoir of potential that still retains essential elements for rewilding deep-soil prairies: generations of ranchers and an intact grazing culture, cost sharing in the form of actual dollars from the state, and support for the long haul from the Nature Conservancy and other groups. The grassland's current level of biodiversity is low when compared to the American Prairie Reserve. Restoring prairies on the scale of the Grand River Grasslands could take generations. As scientists unravel the secrets of fostering delicate late-successional species and weakening invasive species, the prairies of the Grand River basin could rise again. Deep-soil prairies evolved over thousands of years. Time might still be on their side.

The one measure of "wildness" that the Grand beats the upper Missouri on is dams. More than 90,000 dams affect the functioning and connectivity of American rivers and their tributaries. No dam has ever been built on the 226 miles of the Grand River's main stem. This required decades of restraint as the river tormented communities with ravaging floodwaters, especially at Pattonsburg. Driving the Grand River valley, and once in my canoe, I kept seeing Kilroy-style graffiti spray-painted on culverts and weathered signs: "Pattonsburg was here." Eventually I found out that Pattonsburg *was* there, but not anymore.

Since Pattonsburg's founding in 1845 in the flats between the Grand River and two creeks, the triad of streams had flooded the small village more than thirty times. Just as they did on every other prairie river in Missouri, these floods inspired calls to act, starting in the 1930s and 1940s with the Pattonsburg Lake proposal. In the 1950s, the lobbying group MoArk drafted plans for five dams on the Grand River to address flooding in the basin, including one large dam on the Grand near Pattonsburg. Over the next two decades, MoArk lobbied Congress to include funds in the Army Corps of Engineers' budget for these projects. This was the era of federal dam building. By the mid-sixties it seemed inevitable that the Grand would be dammed to prevent catastrophic floods in towns along the river. But people who lived around Pattonsburg opposed the dam, claiming their town would be inundated by the flood control pool to save farmlands downstream. The project was shelved in 1974.

In the summer of 1993, a July monsoon fell straight and heavy, sometimes leaving an inch or more of rain per hour on the soggy ground. Pattonsburg, like so many times before, began to flood. Edwin Howard, who was mayor in 1993, told me the flood was a worst-case scenario for the village. "In previous floods the water moved through and was gone. This time, the valley just filled up like a sink with Pattonsburg at the bottom."[13]

Bar tables at the town's only tavern floated away with open beers. While locals had grown accustomed to floods and could handle the muddy basements, rotten floorboards, and stranded pigs that came with them, the 1993 flood saw entire homes swept into the river and emergency rooftop rescues. When the waters subsided, people came downstairs to find dead animals on living room tables.

After the flood, the city considered its options. The Federal Emergency Management Agency's hazard mitigation program had resources to not only clean up the town but move it to a new site altogether. Compared to MoArk's historic plan for five reservoirs to control the Grand River entirely, moving a town of three hundred (down from a peak population of two thousand years before) was a bargain at $12 million. Howard said, "The process started in 1994 after a series of town meetings."

Gradually a consensus arose to move the town out of the floodplain. "The federal money was enough to move everybody to the new town. More money was raised to build a new school, senior center, city hall, that kind of thing."

A new, environmentally friendly, pedestrian-oriented town was eventually constructed two miles away on a contoured ridge close to I-35. The citizens of this traditionally conservative Missouri village had made a radical choice. They opted to let the river win—which it was destined to—and save their valley. Pattonsburg picked itself up and moved, eventually building not only a city where nobody had to walk more than five minutes to downtown but even a space-aged monolithic dome high school.

The old town was left behind. In 1997, a Hollywood film crew for Ang Lee's movie *Ride with the Devil*, based on Missouri author Daniel Woodrell's novel *Woe to Live On*, used the ghost town to re-create the Lawrence Massacre of 1863, an attack by confederate guerrilla raider William Quantrill that resulted in the burning of much of the city. After the film crew tore down the village for theater, old Pattonsburg really *was* gone.

Over the course of a year, I paddled most of the Grand River in Missouri. There were spots where people had used it as an open dump. Long miles of channelized monotony on straightjacketed stretches made me feel like a pixel following a vector in a 1980s video game. But after more paddling, soon enough two does drinking at the edge of a sandbar would flick their tails when they saw me. Then I'd lean into a few good strokes, glide the curve of a bend, and parade down a riverine avenue of cottonwoods, slender nettles, wild ginger, and the garnet lantern flowers of pawpaw trees whose ancestors might have bloomed in spring forests along the river for five hundred years.

The Grand River is a survivor. Each spring, when thunderstorms migrate east across the prairies, it still floods, unleashing chaos on the bottoms, fighting to carve switches back into the channelized cat-runs of another century, patient in its mission. It is a sculptor of city-sized logjams; a mover of actual cities; a canoe-eating grassland river in a state of famous rivers.

And, as a *true* prairie river, whose entire watershed was once enveloped in a valley of grasses, the Grand was perfect prelude to my next river, itself a river of the bluestem, the longest prairie river in North America. Burned into my circuitry, it was the first river I loved, a sentiment echoed by a growing community determined to spread the word about this once-forgotten natural treasure.

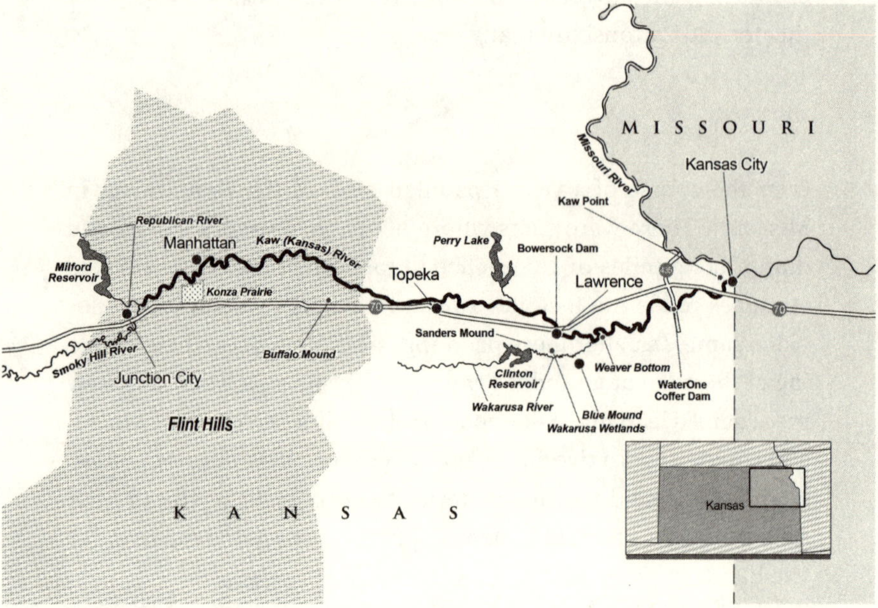

MAP 4 The Kaw (Kansas) River

CHAPTER 4

KAW

At midnight, the world is quiet. From an ancient overlook above Yankee Tank Creek, the high beams of eighteen-wheelers barreling down a highway that outlines the black silhouette of the Wakarusa River appear as beacons in the void. For wandering minds, both the road and the river are narrow and dangerous. On nights like this, spring winds conjure visions of an older highway, one that could connect prairies of the past to prairies of the future. Blue Mound, Shank Hill, the Yankee Tank overlook, Sanders Mound, the red prairie at Mound Cemetery, Burnette's Mound: fill in the gaps with a few unnamed bluffs and you could hike a line-of-sight traverse butte to butte sixty miles west to Buffalo Mound in the Flint Hills. This imaginary promenade traces what archaeologists believe was the last bison migration route in the eastern valley of the Kaw, the longest *true* grassland river in North America.

Measuring 750 miles when the length of its longest tributary is included, the Kaw, or Kansas, River, as listed with the United States Board of Geographic Names, officially begins where the brown waters of the Smoky Hill swirl into the ruddy current of the Republican River just below Junction City: from there it flows 173 miles to its rendezvous with the Missouri in Kansas City. If you want to learn to love a river, 173 miles is an ideal length, almost, but not quite, too long to fully absorb in one lifetime when sipped slowly, bonfires blazing.

The Kaw is a *true* grassland river. It doesn't spring from the melting snowpack of an alpine talus field. The river drains a patchwork of watersheds that contain every conceivable permutation of North American prairie save for the desiccated Llano Estacado semi-deserts of Texas and New Mexico. The Kaw's major tributaries include the Smoky Hill, perhaps the greatest bison incubator of the previous millennia; the Arikaree, whose intermittent waters, which dribble across a loess badlands of yucca, buffalo grass, and prickly pear, are shrinking due to groundwater pumping in the Ogallala Aquifer; the Saline, known to the Pawnee as *Ne Miskua*, the longest shortgrass prairie river in Kansas; the Solomon, a sandy, brackish river of salt springs and mixed-grass prairies; the Republican, named not for the political party but for the Pawnee, who had a western Kansas band called the Republicans; the Big Blue, a tallgrass stream where the Kaw people lived in their Blue Earth Village; and the mysterious Wakarusa, the final great southern tributary before the Kaw joins the Missouri.

Like the Grand, the Kaw is a quintessential tall prairie river. It divides the northern and southern Flint Hills, a three-county-wide swatch of mostly never-plowed tallgrass prairie in Kansas and northern Oklahoma. Outside of the Nature Conservancy's 8,600-acre Konza Biological Research Station, 98 percent of those prairies are in commercial ranches with little public access, save for the riverine highway of the Kaw, one of three officially "navigable," and therefore public, rivers in the state. The National Park Service added the Kansas River to the National Water Trails system in 2012 although, unlike the upper Missouri, the Kaw is not designated wild and scenic. No matter. Thousands float the serene Flint Hills section each summer. Since 2012, outfitters have sprung up; high school environmental studies classes and entire theater troupes paddle the river in great flotillas of colorful kayaks; triathlons of mountain biking, running, and kayaking use the river as a course; and fundraisers offering unlimited samples of beer brewed from the (filtered) waters of the Kaw generate money for river causes. After a century of abuse and neglect, the river has entered the consciousness of ordinary Kansans and, recently, adventurists who come to tackle its entire length in one go. Friends of the Kaw, one of the most well-organized river advocacy groups in America, has been key to the renaissance. Founded to oppose in-channel sand dredging, the group is a member of Waterkeeper Alliance, a global grassroots network that fights to protect clean water.

The Kansas River could be the poster stream for grassland rivers. Because I'd paddled it for most of my life, I thought I knew it well. But after journeys on the upper Missouri and the Grand Rivers, I wanted to revisit the Kaw's final fifty miles. Unlike the Flint Hills, the wet grasslands along the lower Kansas River are ghost prairies. The Wakarusa Wetlands, one of the most beloved wild places in the state, is the best remaining example and a living laboratory where researchers are studying the secrets of wet prairie restoration.

The Grand, however, had repo'd my river wheels. After trolling Craigslist, I bought a used boat and christened it the *Getting to Nowhere*. The lightweight solo was perfect for grassland rivers and ergonomic enough for the MR340, the ultramarathon river race that loomed on my paddling horizon.

I also had an obscure written inspiration for the journey. The journals of Tom Burns, the last commercial fisherman of the Kaw, describe a fantastical migration that seemed almost too strange to believe.[1] By paddling Burns's route at the same time of year he described, I hoped to witness the phenomenon myself. With or without the company of migrating beasts, however, the trip would help me accomplish a personal goal that was embarrassing to reveal. The power of the lower Missouri River had always frightened me, so whenever I paddled the Kaw in Kansas City, I took out at a fishing dock to avoid the confluence a mile downstream. In all my years on the Kansas River—*my* river—I'd never paddled its final mile.

The biology of catfish is a murky science. Prairie waters are home to dozens of species, ranging in length from finger-sized madtoms of the genus *Noturus* to cucumber-sized bullheads of the genus *Ameiurus* to the big-three game species—channel catfish (*Ictalurus punctatus*), flatheads (*Pylodictus olivarus*), and blue catfish (*Icta-*

lurus furcatus). All of them share characteristic barbells or "whis-kers" that inspire a feline nomenclature. Flatheads, or yellow cat-fish, are the tigers of the group. Solitary and nocturnal, they lurk in the cover of underwater snags and rocks to ambush prey, feeding mostly on live fish. Flatheads fight each other and protect home territories to the death. For most of the year they stay within a one-mile radius of their nests. Catfish have poor eyesight, but ichthyol-ogists believe they use densely packed taste buds on their barbels and skin to build olfactory maps of amino acids, helping them nav-igate upstream. In the last half century, anglers have reeled in cat-fish weighing up to one hundred forty pounds, though legends of much larger fish persist from earlier years.

These tales of mythic catfish were part of the world Barbara Higgins-Dover grew up in. When she was a child in the 1960s, she knew that her family wasn't "normal." They were commercial fishermen. Today she curates Kansas Riverkings, a small museum that commemorates the commercial fishing era on the Kaw. Her grandfather, Richard Higgins, owned a fish market on the north side of Lawrence near the river. They hung catfish larger than her-self on clotheslines in the backyard and peeled the skins off with pliers, making a sound she told me was "super creepy."[2] Her fa-ther inflated their fragrant bladders for her to play with. What she didn't understand then was that her grandfather's craft was part of a nearly extinct tradition that had begun a century earlier. Jake Washington, born into slavery in Missouri in 1849, was one of its founders.

The details of Washington's early life are uncertain, but Higgins-Dover told me that from her research she believes he escaped en-slavement and followed the underground railroad from Quindaro to Lawrence, an abolitionist stronghold prior to the Civil War. People escaping slavery were in legal limbo before the Emancipation Proc-lamation, and many were forced to inhabit the margins of society. Bottomland forests north of the Kaw served as a refuge. For a while, Higgins-Dover told me, Washington made a living selling wood for lumber and the furnaces of steamboats traveling upriver. He also learned to fish with hoop nets. Different nets trapped different species of fish depending on the width of the mesh and the depths where they were placed.

After the Civil War, Washington met Abe Burns and taught him everything the much older Washington knew about the river and fishing. Burns quickly mastered the art and developed a new method that made the duo famous. The Kaw was a free-flowing river until the Bowersock Dam was constructed in 1874 to provide mechanical power for small industrial plants. The dam completely blocked the river, stopping large catfish moving upstream from the Missouri and creating a deep mill pond. Legions of catfish roamed the depths between the dam and river holes near Mud Creek six miles downstream. This concentrated fishery ushered in a golden age of commercial fishing on the Kaw.

Burns's fishing method was spectacular. He and Washington would arrive at the dam before sunrise and set their hoop nets. Then Burns would prepare to fish. He would kick off his boots, take off his shirt, lash an iron gaffe around his right arm like a pirate's hook and start taking deep breaths. When he was ready, Burns would dive headfirst into the millpond below the power plant. The hydraulics beneath low-head dams like the Bowersock create centrifugal death traps, black holes that trap cottonwood trees, boats, and unfortunate souls. Because surface currents sweep trapped objects back toward the dam, escape is nearly impossible. The only way out is to dive to the bottom of the river, where the centrifuge might spit you downstream to safety, if you could hold your breath long enough.

Burns could hold his breath long enough. The brown waters of the Kaw were opaque, so after he dove in, he blind-waved the gaffe hoping to bump into a giant fish. If he noodled one before he ran out of breath, both fisherman and fish would be swept downstream like a waterslide. Word of the duo got out. In the 1880s, fifty or more onlookers would gather daily to watch from the bridge. Abe and Jake, as they were known, became legends.

Soon other fishermen joined them. Most fished with hoop nets and set lines. They built three miles of fishing huts to the east of Lawrence and learned the river by watching one another, stealing tricks, reverse-engineering discarded gear, and vying for customers. Their lanterns shone like fireflies on June nights. Herons and crows would crowd near to steal bait. Some of the men brought their families and developed side hustles, renting boats for joyrides and setting up community ice-skating rinks in winter. For a quarter century these anglers developed a deep knowledge of the river and its fauna.

In the early twentieth century, times got tough for the fishermen of the Kaw, in part because of Lewis Lyndsay Dyche. Dyche was a naturalist, buffalo hunter, explorer, and the taxidermist who stuffed the horse Comanche, the only survivor from the US Cavalry forces at the Battle of Little Bighorn. In 1909 Dyche became the state's first fish and game warden. He oversaw construction of a state fish hatchery and formalized Kansas's loose hunting and fishing laws, bringing them in line with regulations in eastern states. In 1911, the "Dyche Bill" was passed and became the beginning of modern Kansas hunting laws.[3] A provision of the bill made commercial fishing on state rivers illegal. Higgins-Dover told me that part of Dyche's motivation was to generate state revenue by selling fishing licenses to the public. In the early twentieth century, the only way to serve fish for dinner was to catch it yourself or buy from a local fisherman. Abe and Jake and their disciples working from cabins along the Kaw were suddenly in the way of Dyche's plans. By 1920, the state had shut down the cabins, and commercial fishing on the Kaw went underground. The river men became outlaws.

Higgins-Dover's grandfather was one of the last of those outlaws. The elder Higgins held a commercial fishing license on the Missouri River, a subterfuge that let him run his fish shop with some level of impunity, even though he sold fish illegally caught from the Kaw. He was arrested three times, went to jail, and had his equipment confiscated. It never stopped him though. He always returned to his catfish runs.

Higgins-Dover believes the river below Bowersock signified both peril and possibility: "The stretch of the Kaw that flows over the dam is dangerous, but beautiful and serene at the same time. It's bipolar. It provides sustenance and food but is also a danger to the

FIGURE 4.1 Jake Washington (left) and Abe Burns with two large catfish caught in the Kansas River, 1896. Photo courtesy of Kansas State Historical Society.

community. I think my grandfather and these men respected this. They learned things that nobody alive today understands. They knew the danger and that's why they thrived along the river."

Higgins certainly knew of its dangers. When somebody went missing in the river, authorities would consult him to search for the body. He could predict where it was likely to be found by the way the river was flowing and whether the dam was letting water spill across the top. Higgins himself was not immune to the river's cruel neutrality. In 1950, he was fishing beneath the dam with his eight-year-old son, Charles, when the boy walked across a slick mossy spot and fell into the river. Higgins couldn't rescue him, and his son drowned.

The stories of Abe Burns and Jake Washington have faded into a misty past. A few newspaper articles exist, including a story published when Washington's son Turtle was hit by a train and killed on the Santa Fe tracks adjacent to the river. But one vivid artifact remains: an 1896 photograph of Abe and Jake holding two enormous fish on a stringer hoisted between them. The 90- and 110-pound catfish are probably flatheads. In the photo, Burns is wearing shirt sleeves, suspenders, and a conductor's hat. He might have been coming from work at the railroad. Washington looks more relaxed with hand on hips and cap folded back. Although the image was cropped, the fish were probably suspended from a rope thrown over a cottonwood above the two. A sculpture based on the photo in "Abe and Jake's Landing," an events venue on the south bank of the river, commemorates their decades-long partnership.

By 1970, the commercial fishing era had ended, but a few masters continued to work the river until late in their lives, including Tom Burns, the last commercial fisherman of the Kaw. After he hung up his nets for good in the 1990s, Burns (no relation to Abe Burns) developed a cult following. One of his devotees was Ned Kehde, a respected Midwestern outdoors writer and contributor to *In-Fisherman Magazine*. Kehde had met hundreds of professional sports anglers and recognized that Burns possessed knowledge about fish and rivers that nobody else did. Burns and Kehde became friends, and through his writing, Kehde introduced Burns to new generations who were as fascinated by Burns's mastery of river ways as his ability to haul in fish.

Tom Burns fished the river almost every day for sixty years, most of the time after working his full-time job. Born in Lawrence in 1919, he developed a sixth sense about the Kaw, Kehde told me, and could read the composition of the riverbed by perturbations on its surface, whether there was a boulder or a log or a mud ball down in the murk. What he really knew were big fish: flatheads and blue cats, gar, shovelnose sturgeon, and paddlefish—the Pleistocene megafauna of prairie waterways. Instead of working from a cabin, Burns traversed a clandestine, nocturnal cat run of fish traps along the river. He sold catfish, buffalo fish, and drum on a black market to Kansas restaurants and rough fish connoisseurs.

Kehde encouraged Burns to write a memoir about his passion for the river. He wasn't a writer, but when he sat down with pen and paper, Kehde said, Burns wrote like Jack Kerouac and "the stories boiled up in this head like a thunderstorm with words pouring down as fast as he could move the pen."[4] In a brief expositional explosion, Burns completed a ninety-seven-page self-published memoir, *60 Years on the Kaw*.[5]

Kehde said that toward the end of Burns's life he agreed to meet with an editor of *In-Fisherman Magazine* to talk about the behavior of flathead catfish. Kehde said the young writer was "absolutely spellbound by Tommy's knowledge of the river, and all the surroundings, and Tommy's ability to 'be one' with it." The editor had never met a man so "physically and psychologically attached to a river" as Burns was, Kehde said.

Ready to test my new canoe, I decided to put in near the haunts of the commercial fisherman at Bowersock Dam, continuing along Burns's section of the river between Mud Creek and Stranger Creek, and on to Kaw Point where the river pours into the Missouri, with Burns's journal for inspiration and guidance.

The night before the trip, I walked to a grassy square near the south foot of the girder bridge that spans the Kansas River in Lawrence. The park was known for a twenty-eight-ton red quartzite boulder. The Kaw people called the glacial erratic *Iⁿ'zhúje 'Waxóbe*: the prayer rock. For millennia it sat near the south bank of the river at the mouth of Shunganunga Creek east of modern Topeka. Many members of the Kaw Nation still consider it sacred.

In 1929, a Topeka attorney wanted to move the rock to the grounds of the Kansas State Capitol. Hearing about this, a group from Lawrence convinced the Santa Fe Railway to help them seize the boulder and bring it to Lawrence ahead of the city's seventy-fifth anniversary celebration. They placed the rock in Robinson Park, named after Charles Robinson, the first governor of Kansas and founder of Lawrence, and mounted a plaque on the boulder with the names of the city's first, mostly white, settlers. It read, "To the pioneers of Kansas who in devotion to human freedom came into a wilderness, suffered hardships and faced danger and death to found this state in righteousness." The plaque made no mention that the Kaw people had lived in that *wilderness* for centuries, or of the rock's significance to their nation.[6]

A storm was gathering in the west and the sun fell beneath the cumulonimbus. Bony fingers of heat lightning flickered in the distance. Beside the prayer rock, and two hundred feet from where Abe Burns dove for catfish, I looked out toward the forest where Jake Washington hid during what must have been a terrifying period of his life. All that remained of the two now was a photograph. The holy rock, a corporeal reminder of the Kaw Nation's centuries-long relationship with the river, was inscribed with the names of the *wrong people*. That night I hardly slept, my fears and insecurities whirlpooling in the collective mind of the river. Who was I to think I *knew* it?

Christina shook her head as she helped me load the canoe at the edge of the stilling pool below the Bowersock electric plant. "Glad you're going without me," she said. June in Kansas can be as humid as a rave in a hothouse, but this was worse. The previous night's storm never materialized to cool things off and the forecast predicted consecutive hundred-degree days. We inventoried my eight vitamin waters, six half-liter alkaline waters, and two extra one-gallon jugs for cooking. To avert hyponatremia, a potentially fatal loss of blood sodium that paddlers can develop from profuse sweating, she pushed five sticks of a dubious looking "vegan jerky" into my fanny pack. "Yep," she said, looking at the limp globs of not-meat, "glad you're going without me."

With one leg in the boat, I pushed off and slung my other leg over the gunwale. Before turning the canoe downstream, I paddled alongside the concrete structure that held the Bowersock plant's retro-futuristic-looking turbines and then along the face of the dam itself. Pivoting back into the current, the reflections of cumulus clouds were spilled popcorn on the glistening water.

Because the river flows across loaves of sand braided together like challah bread, finding the channel is constant entertainment for Kansas River paddlers, especially at low flows. But summer snow drifting down from the cottonwood canopy made a convenient trail to follow. I paddled slowly to get to know the new canoe, and the first miles where the old river men had worked were pleasant and relaxing. Near a shallow flat, a set of wooden steps led to nowhere at the top of the south bank, a relic, perhaps, of one of the fishing cabins. Pileated woodpeckers and a kingfisher called from the forest.

While camped in the Flint Hills in 1806, Zebulon Pike wrote in his journals that in a single view from a bluff looking down onto a river bottoms he saw "buffalo, elk, deer, cabrie, and panthers."[7] ("Cabrie" was a nineteenth-century nickname for antelope that we should resurrect.) The current took me past a clearing on the north levee where I once looked out over the floodplain and counted 103 white-tailed deer, one fox, two raccoons, and one coyote that used its snout to flip a skunk over onto its back. *Take that, old-timers*, I had thought to myself then.

Fisheries biologists long believed that flatheads were solitary homebodies who kept close to their lairs in submerged logjams deep in the river. But newer studies in the Missouri and Mississippi Rivers show that not all flathead populations behave in the same ways.[8] When water temperatures drop to about fifty degrees, some fish travel as far as sixty-five miles into tributary streams where they overwinter in deep pools. When spring comes and water temperatures rise, the big muddies continue upstream to spawn, then return to their summer spots. They can be half a state away.

Tom Burns discovered this himself fifty years before biologists proved it by implanting radios in fish. In his memoir and an *In-Fisherman* interview with Kehde, Burns said that flatheads migrate in schools, traveling about five miles per day upstream.[9] The average mature Kansas River flathead, he said, weighs more than twenty pounds, so a school with hundreds of fish would look like a pod of bonsai dolphins in sandy shallows. Burns discovered that hundreds of flatheads made an annual migration to a deep underwater canyon near Mud Creek. In the coldest winters, all the oldest, gnarliest flatheads in a hundred miles of river gathered there, one of the most prolific wildlife concentrations in Kansas. Mud Creek was his "honey hole," where even in winter he caught catfish using a fifteen-foot pole fitted with a dagger end to spear the torpid fish through holes he chopped into the ice with a hand axe. He set hoop nets above the hole. Some years, he blocked as much of the flathead migration up toward Bowersock as he chose to. It was a magical spot for Burns. He wrote that on moonlit nights, he heard coyotes and owls answer the train crossing Mud Creek as it blew its warning: two long whistles followed by one short whistle and then another long whistle.[10]

One day, after the flathead migration had begun, Burns fished from a rock next to a sandbar with deep holes on either side. The water was smooth and placid when he arrived. Then, in an area the size of two football fields, hundreds of enormous flatheads swam up and began spinning around and "playing" with each other on the surface of the river. This was the strangest thing Burns ever witnessed, and he only saw it once. Kehde said Burns didn't divulge the exact location, but it was somewhere between the deep hole at Mud Creek and Stranger Creek sixteen miles downstream. Kansas Department of Wildlife and Parks fisheries biologist Ben Neeley, who has studied the behavior of flathead catfish in the Kaw, told me he had never seen such a mass migration himself, but it was possible given the right conditions. "Flatheads certainly exhibit long-scale movements," he said. "It happens about the time they reproduce, and they all do this at the same time. The dams create issues for them in the Kaw, especially the Bowersock that blocks all upstream movement for flatheads in the lower river. The Kaw has a braided sand channel, and once the water gets low there aren't many places for the fish to go. If Burns was in a spot with one passage and it was the day the fish were moving through, they could have all been 'stuck' together in the same place at the same time."[11] It was a long shot, but I hoped to get lucky and witness rolling flatheads in the hours before sunset.

The sandbar at Mud Creek had grown up in the twenty-five years since I first paddled the river. Cottonwood saplings had colonized a campsite where I had taken friends years before. Using J-strokes on the starboard side to steady the boat, I paddled up to the old spot and remembered those irretrievable nights of my youth—the damp chill of a June evening, sitting in darkness by the inlet with my friends, splatting mosquitoes while waiting with flashlights for families of beavers to swim up to a buffet of young cottonwoods; a shimmering night on the sweet sand with a girlfriend, a sudden heat storm striking at midnight, huddling in the tent while bursts of lightning popped like trays of flash powder, then laughing and tumbling out onto the wet sand cool against our skin, the half-moon fighting through dissipating storm clouds to illuminate our young bodies.

The river too had changed since those days. After the 1993 flood, the Kaw carved a new channel across from the old sandbar, creating an island of sycamores, hackberries, and cottonwoods. I lifted my paddle from the water to list above one of the deepest scour holes in the river, at least seventeen feet this day, and much deeper in high water. This was probably Burns's honey hole. From here to Stranger Creek, I'd be on the scout for dancing flatheads.

Below the island, cottonwood forests surrounded the river. Creeper vines and poison ivy smothered the tallest trees, balding the bark from some. In his book about the Kaw Valley, Jim Locklear writes: "I've seen many glorious cottonwoods, but some remind you of running into that old friend who has lived a very hard life or maybe made some very bad choices."[12] The swoon of overgrowth was making all the choices for these old titans.

More vegetative carnage soon followed. In his journal, Burns writes that he was blown out of his rowboat once by a tornado. A quarter-mile-wide wound on the right riverbank marked the path of the 2019 F4 tornado that crossed the river twice and devastated farms in the floodplain. Scars from the gash will remain for at least a decade. A breeze pushed me toward the gash, but I fought back with the new canoe to stay midchannel. Below the Eudora bridge, the river turned north into Weaver Bottom, bare farmland that once was a deep old-growth forest and, later, a village that was abandoned due to repeated flooding.

Ready for a break, I tied off at a limestone ledge and set up my camp chair on the sandbar to read. Burns's ninety-year-old mother lived on a farm near Weaver Bottom. One morning in 1969 she went to her barn to see why her dogs were barking. One reared up on its hind legs, and she realized it was a black bear. Bears had disappeared from Kansas in the late 1800s. This one was probably a young male wandering north from the Ozarks in southern Missouri. That night, according to his memoir, Burns fished near where I was sitting and at 3:30 a.m. heard what sounded like a man calling for help. He rowed his dory toward the sound. It was the bear howling. He built a bonfire and sat up until sunrise. In the morning he found the tracks and followed them into the farm field until they disappeared.[13] Before 1850, black bears were common along the Kaw. Burns might have witnessed the only bear to wander the Kansas River bottoms since the Kaw people hunted them in cottonwood groves a hundred and fifty years earlier.

I had just returned the memoir to the pack when a fish torpedoed through clear water below the rock ledge. It surfaced when it reached the shallows, and just like Burns described in his fable, rolled over and exposed its smooth white underbelly and barbels. A flathead! I trained my binoculars on the green narrows hoping it was the start of a migrating school. Any fish moving through would have to pass this spot since the water farther out was no more than two inches deep. The big fish splashed in the shallows for a few minutes and continued upstream. For the next hours, I chewed vegan jerky and watched catfish shoot the narrows one at a time, each stopping to splash around. Midway between Mud Creek and Stranger Creek, this could have been the spot where Burns saw the migration.

But by 7 p.m., the water, where two dozen flatheads had barreled upstream one by one, had darkened in the fading light. I edged the canoe back into the water, hesitant to leave on the chance the next big fish would lead to a hundred more. The annual spring passerine migration along the central flyway was almost complete, but I'm sure a few birders had worked the cottonwood bottoms of the Kaw during the day. Had I been the only *fish watcher*? Was that the correct term for a fisher who didn't angle? Reading Burns taught me a new way to see rivers and think about the strange animals living in the waters of those rivers.

I reached the mouth of Stranger Creek a half hour before sunset and beached on an annex of sand unreachable from shore. I didn't want to camp there, because the highest point was only four inches above the river. There were two perfect sandbars farther downstream below a high rocky bluff. Four tents were clustered around a fire on the first sandbar, and three men with a fishing boat were gathering driftwood on the second. The downmarket sand island would have to work. If the Corp of Engineers released water from Tuttle Creek during the night, I'd have to swim for it. I unpacked my gear, built a fire ring of sycamore knots, and snapped my camera on its tripod in the small chance hundreds of migrating flatheads might decide to swim by. Instead, I had Burns's writings. On the last page of the memoir he wrote, "Now when the shadows grow long . . . [I] take a moment to think of the joy those nights held for me during those wonderful hours on the Kaw, tangling with its beasts."[14]

Taking a beer out of my cooler, I popped the cap and raised a toast to river beasts and Tom Burns, the *king*, as they called him, of the Kaw.

It was a long, sweaty night. Hurrying to set up camp, I hadn't shaken all the sand off my sleeping bag and pad before throwing them into the tent. Tossing around trying to find a comfortable sleeping position, a thin, exfoliating batter soon coated my chest and neck. It was my kind of spa treatment, a gritty reminder of how precious river time was. Over the years, sand from the Kaw has found its way into my toothbrushes, belly button, and ears. I have found Kaw River sand coating the inner tubes of my bike tires, along the baseboard of a bed in an Airbnb we'd visited two years earlier after a river trip, on the roof of my house and in the roof of my mouth, and on the carpet of a conference room below my feet in India. I've noticed Kaw River sand in the shape of a kiss on my wife's cheek the morning after I returned from a weeklong float, in my cat's downy inner

FIGURE 4.2 The Kaw River near the Rising Star boat ramp, Lecompton, KS. Photo courtesy of Lisa Grossman.

coat, and in my daughter's crib when she was a baby. I have found it in unopened bottles of beer. A friend was pretty sure she saw some on a sonogram of her abdomen. After I die, if anybody looks, they will find it mixed with my ashes.

Besides water, from a strict economic viewpoint, sand is the most valuable product of most grassland rivers. The sands of the Kaw originated millions of years ago in boulders and rocks that eroded out of the Rocky Mountains and were carried east by streams that ground larger pieces down into subsequently smaller and smaller grains. The golden granules of sandbars in active river channels and buried off-channel deposits are more than metaphorical to the sand mining industry. This has created tension between sand mining companies and environmentalists.

Sand is one of the main sediments transported by prairie rivers. The relationship between hydrology and sediment load is a complicated one. Stream velocity, depth, and the slope and width of the channel are variables that determine whether sediment gets carried downstream or dropped into semistable deposits. The Kansas River carries 1.5 million tons of sand per year in mean flows, although that can vary in droughts or flood years. I asked Heidi Mehl, the director of Kansas Water and Agriculture Programs for the Nature Conservancy, about the geology of sand in prairie rivers. She told me that in the same way Kansas is prone to tornadoes because of its location between cold dry air masses from the west and warm humid air masses from the Gulf of Mexico, the Kansas River is located an ideal distance from the slowly eroding Rocky Mountains for the Kaw to grind the ancient mountain sediments into granules that are the perfect size for construction applications.[15]

Mehl explained, "Because the Kaw is a completely prairie-based river with a watershed that doesn't extend into the mountains, it doesn't get gravels pushed downstream by high mountain flows. Big flushing flows coming down from the mountains affect sediment sorting. Instead, the Kaw recruits sands that were washed down millions of years ago and are now held in the prairies of western Kansas, eastern Colorado, and southern Nebraska. The balance over the geological ages has favored sand or put enough sand in the system that that's the majority grain type. Because the river has few major dams on the main stem and even on nearby tributaries, development hasn't ruined the sands of the Kaw."

On grassland rivers, sand can be mined either by hydraulic dredges in the river itself or from pit mines that tap into ancient channels of the river valley. The Army Corps of Engineers is the organization that grants sand mining permits in the United States. Pit mines require more land than river dredges and compete with other users of the floodplain, like farms and the large rural homes of city people who flee to the countryside. This increases the cost for pit mines and can create local resistance to new permit requests. Jerry Younger, managing director of the Kansas Aggregate Producers Association, the trade group that represents companies that mine and produce aggregate products like sand, told me that it gets harder every year for producers to obtain a permit for a new pit mine.[16]

In-channel mining requires no land for the dredge itself—the Kansas River is a public resource. This reduces costs. Mehl said the river also does some of the work for sand miners. "Since the river is always flowing," she noted, "it's going to continually sort the sediment, so it's easier to pull the grain sizes used for construction. The sweet-sized grains for commercial use are the same size that sandbars are composed of."

Younger told me that all river mining operations are monitored. He said, "Every two years riverbed elevations have to be surveyed, and the producers pay for the work." The surveys monitor how much the riverbed is degraded by the dredge. The Corps of Engineers analyzes each survey to determine whether the permit can continue for another two years. As of 2024, permit holders pay fifteen cents for every ton of sand they remove, a price that hasn't changed since 1996.

Sand extraction has an environmental cost, especially in-river extraction. Hydraulic dredges dig up not only sand but also chlordane-contaminated sediment from the bottom of the river. Mehl said that dredging can damage the riverbed, destabilize stream banks, affect water quality, aquatic life, fish spawning beds, and, by increasing sediment load, make it more expensive for water purification plants to produce drinking water. Lateral erosion can occur when the river fills dredge holes by taking sand from the banks above dredges. This can cause banks to collapse and swallow strips of riparian forests and other streamside vegetation. Because heavier particles typically settle into dredge holes, lighter sediments will separate and float downstream, creating thick layers of silt that choke out aquatic life.

I was glad for the fine-grained sediment composing the little island. The sand made a plush mattress beneath my tent. But the Stranger Creek sandbar had other problems when it came to sleeping. Long trains clattered by on the Union Pacific tracks like tools of psychological warfare. Swarms of mayflies had hatched under the light of the full moon and a few hundred forced their way into my bedding. In the past, snowplows were used to clear mass emergences of *Hexagenia bilineata* from bridges across the Mississippi River. Lying awake in my tent, I cursed not mayflies but my timing. The week before I had been invited to join a group float on this same section of the river. If I had joined them, I might have been spared the mayfly inferno.

Friends of the Kaw organized the float I had been invited to join. The Kansas-based river advocacy organization can trace its roots back to conflicts over sand. In 1991, a group of neighbors who lived near the river were sick of the constant noise from dredging, heavy machinery, and truck traffic that an on-river sand mining operation was creating near their homes. When another permit was presented to the Army Corps of Engineers, they banded together to fight the application. Since then, Friends of the Kaw has become one of the most successful river advocacy groups in America. They've mounted public opposition to stop sand dredging permits on the most pristine sections of the river. After founders Lance Burr, Laura Calwell, Mike Calwell, and Charles Benjamin unsuccessfully lobbied the Kansas legislature to pass a bill to halt all new dredging permits in the early 2000s, they convinced Governor Kathleen Sebelius to create a task force to study the best way to move sand mining off river. Since then, only one new in-channel permit has been approved, and the Army Corps of Engineers has shut down other permits that caused unacceptable riverbed degradation.

Friends of the Kaw employs a salaried "Riverkeeper" to monitor and mediate sand mining operations and suspected pollution incidents and to advocate for the river. The idea for such a role wasn't new. It was started in New York's Hudson River valley by a group of blue-collar fishermen who joined together to fight industrial polluters. Today, this "Waterkeeper" movement is an international federation overseen by the Waterkeeper Alliance, a group of three hundred Waterkeepers around the world who, according to the group's mission statement, are "on the front lines of the planetary environmental crisis, patrolling and protecting more than 2.5 million

square miles of rivers, lakes, and coastal waterways on six continents."[17] Waterkeeper groups oversee rivers in the People's Repub-
lic of China, Colombia, Nepal, Peru, Senegal, Canada, the United
Kingdom, the United States, and many other nations.

Dawn Buehler became the fourth Kansas Riverkeeper in 2015.
She grew up on her family's vegetable farm along the river near De
Soto. When New Wave music was playing in 1980s suburban shopping malls ten miles from their white farmhouse, Buehler inhabited a different world. Her father, she told me, believed in the power
of the forest. "He had no formal training as a conservationist, but
he knew the deep-rooted trees protected our home from the river.
The canopy in those old-growth woods allowed natural light onto
the floor, and every spring we'd go hunting for mushrooms. It was
deep and beautiful."[18]

After a canoe trip in her teens, Buehler was "110 per cent sold
on the Kansas River. Being here with nature, at the same level as
the river, was an intimate experience for me. Complete freedom."
By the time Buehler joined Friends of the Kaw as Riverkeeper,
she'd acquired an intimate knowledge of the river after decades of
paddling it.

Mark Dugan, an attorney who served as president of the organization's board, told me that under the Clean Water Act, individuals
and groups require a permit from either the Environmental Protection Agency (EPA) or a state agency such as Kansas Department
of Health and Environment (KDHE) to discharge into a river. "If
somebody violates their permit, or doesn't have one," Dugan said,
"KDHE or the EPA can attempt to enforce it. If they don't, individuals or environmental groups can."[19] Friends of the Kaw reports
permit violations to KDHE and the EPA, but Buehler is a mediator
at heart. She's used her reputation as a fair-handed authority with
a deep understanding of the intricacies of water policy and river issues to keep incidents out of court.

As Riverkeeper, Buehler chartered a top-notch team of river
guides—scientists, professors, and seasoned paddlers—to accom-
pany her on "river checks." Industrial discharge is regulated by the
National Pollutant Discharge Elimination System under the Clean

Water Act. In Kansas, KDHE issues discharge permits. Friends of the Kaw maintains a database of those permits, and Buehler and the river guides float the entire river each year looking for unauthorized pipes and effluents. The guides also double as security. She's received threats. Once a landowner told a local fire chief that if the group crossed onto his property during a tire cleanup, he'd "shoot every one of them dead."

Friends of the Kaw, and Buehler in particular, is better known for getting people out on the river. Since the group's formation, paddling has become a multimillion-dollar industry in eastern Kansas. Friends of the Kaw and the Kansas Department of Wildlife and Parks have constructed fifteen concrete boat ramps along the river, creating an access point near every town between Junction City and Kansas City. It effectively "opened" the river. This was one of the reasons the Kaw was added to the National Water Trail system in 2012. The Kaw River guides take hundreds of people out on the river each summer. Buehler believes sandbars are the stars of the show. As wild freeform campgrounds, the sandbars of the Kaw are the best-kept secrets in the entire American backcountry.

Late in the night, strange sounds woke me. Dolphins—at least what sounded like "Flipper" from the perpetual television reruns of my youth—were circling the sandbar. With hushed fingers, I lowered the zipper of the inner vestibule, then the rain fly. As mayflies streamed through the opening like people pushing into a subway, I stepped out onto the sand. The dolphin sounds stopped. Dazed with sleep, my eyes slowly adjusted to the yellow moonlight. After a moment, I heard three distinct splashes, presumably the dolphin-sound-makers plunging into the river. Embers of bonfires on all three sandbars glowed with an eerie translucence. River otter tracks circled the tent like a racetrack.

River otters are a wildlife success story in Kansas, Missouri, and other prairie states. Extirpated from Kansas since 1904 because of declining beaver populations, a species that otters depend on for den sites, otter populations have rebounded after reintroductions beginning in the 1980s. In Kansas and other prairie states, floodplain wetlands and adjoining wet prairies are a crucial resource for the semiaquatic members of the weasel family. Wet prairies were once common along the Kaw and its tributaries. Today, they are one of the rarest types of grasslands in the Midwest.

Prairies can be classified into three moisture categories: dry, mesic, and wet. Dry prairies are found on well-drained hilly up-lands and slopes. Their soils are poorer for agriculture, so much of the remaining unplowed tallgrass prairie in America, and the Flint Hills in particular, is dry prairie. Mesic prairies hold more mois-ture. They were choice pastures and most of them were plowed. Wet prairies are like mesic prairies, but their soils are sandier, have fewer nutrients, and are often found near rivers. During floods, they reduce peak storm flows. "Wet prairies produce a spongy effect that slows down the water and holds onto it during hard rains," Mehl, the geologist, told me. "It reduces erosion in the channel and has benefits for aquatic life and for the sediment regime. Land that's been plowed for agriculture where aggregate stability isn't very good doesn't hold rainwater the way wet prairies do."

The problem doesn't stop once rainwater, and the sediment it carries, hits the river. Mehl added, "Instead of that wet prairie sponge and the elongated lower storm peak, you have a larger quan-tity of water reaching the channel all at once. That's where we get incised channels, bank instability, and mud being recruited from the bank and going into the river and possibly covering sandbars."

Earlier in the day, I'd paddled past the mouth of the Wakarusa River, the largest tributary of the lower Kaw. In pre-European settlement times, the Wakarusa Valley was a tiled mosaic of wet cordgrass prairies—more than eighteen thousand acres in all. During the last glacial epoch, an ice dam formed on the Kansas River. Water diverted by the dam engorged the Wakarusa's valley, creating an enormous floodplain—and a river that geologists call a "misfit stream." The plain was dotted with wet prairies, but today only about one thousand acres remain. The rest were converted into soybean fields, freeways, and the rapidly silting bottom of Clin-ton Reservoir, a federal flood control lake outside of Lawrence that altered natural flooding of the river.

The valley has a long cultural history. The Kaw people lived in villages near modern Eudora and a prominent prairie hill called Blue Mound. According to an ethnohistory of Native Americans and the Wakarusa River valley (part of the environmental impact statement for the South Lawrence Trafficway, a highway built in the bottoms next to the river), the Pawnee, who lived on the Republican River in northwest Kansas in the seventeenth century, traveled to Blue Mound in the springtime for gatherings.[20] Daniel Boone trapped beaver along the river between 1807 and 1815. A few buffalo and herds of elk could be found in the area well into the late nineteenth century, at the eastern terminus of a buffalo migration route through the Wakarusa valley. According to geographer Jay T. Johnson, the word "Wakarusa" is so old that nobody knows for sure what people's language it comes from, whether Kaw, Osage, Potawatomi, or an earlier group.[21]

Today, the Wakarusa Wetlands span 927 acres, 80 percent in restored marshes and wet prairies, 15 percent in floodplain forest along the Wakarusa, and 5 percent in two virgin wet prairies. In February of most years, during the first cold rain, thousands of small-mouth salamanders crawl out of the wetlands at night in a spectacle that closes roads. Muskrats, beavers, and river otters are common in the canals and lily pad pools south of the highway and, according to data compiled by Baker University, 278 species of birds, 98 vertebrate species, and 487 plant species have been cataloged in the preserve.[22]

Today, Haskell Indian Nations University and Baker University are working to restore the Wakarusa Wetlands. Baker's work is led by Irene Unger, wetlands director and biology professor there. A plant ecologist with a PhD in forestry, she has a lopsided passion for prairies or, when she's joking, "weeds." Her doctoral research was focused on the effects of flooding and soil properties, including microbiota, on prairies. Before most of it was plowed in the 1850s, the wet prairies of the Wakarusa Wetlands were subject to semiannual flooding, so her research prepared her for the kind of restorative work she oversees today. She told me, "From our experience, you can get a lot of plants back through a combination of overseeding and burning to release the seedbank that's still there. Maybe not all of them, but a lot. What's different is the underground microbial community."[23]

As with deep-soil prairies in the Grand River watershed of northwest Missouri, "unseen" subsurface ecology could play a crucial role in effectively restoring all types of grasslands, including the wet prairies of the lower Kansas River basin. In a broad sense, Unger told me, soil microorganisms are important for nutrient cycling—they break down dead plants and animals and make nutrients available for other plants and the animals that eat the plants. "Different soil microorganisms can break down different organic molecules. Most, if not all, can break down simple sugars and amino acids. However, as the organic molecules get more complex, fewer microorganisms can break them down," she said. "For example, cellulose and lignin are large structures with lots of carbon rings that have carbon-carbon double bonds; relatively few microorganisms have the enzymes necessary to break these rings. So, if you don't have a diverse soil microbiota, then you might not be able to break these complex molecules down to their component parts."

On a finer scale, at the "rhizosphere," the halo of soil directly surrounding plant roots, microbiota can vary from species to species. Unger said, "Plants influence their rhizosphere through chemicals that the plant releases into the soil called exudates. These exudates will have different types and amounts of carbon compounds. The exudates can contain allelopathic chemicals—chemicals that influence target organisms and can act like toxins. Thus, plants can inhibit the growth of other plants around them. The exudates can also alter the microbial community around plants, either helping the plant maintain its space by preventing the establishment of another nearby plant or helping other plants move into the space."

So, what came first: prairie plants or their microbiota? Can one exist without the other? The question is merely academic for unplowed prairies but could be essential for prairie restoration. Unger isn't sure yet: "We need a robust microbial community for the basic functions of nutrient cycling, but we also know that plants can and do influence microorganisms in the soil. And disturbances such as plowing, fire, grazing, and drought affect both plants and microbes.

It's the chicken or the egg question of grasslands, and we still don't know the answer."

The science of tallgrass prairie restoration has progressed quickly in the last fifty years, but the effects of that science might take longer to realize. "Rejuvenating prairie microbiota is taking longer than we expect it would," Unger admitted to me. "This seems counterintuitive. Microbes can reproduce quickly. You'd think they would respond when new plants come in, like in a restoration."

To some extent, they do. For example, if you plow a prairie and plant crops, the microbiota changes right away (much of the prairie microbiota dies). But when you rehabilitate soil and bring back prairie plants, Unger said, it seems to take longer for the microbial community found in an unplowed prairie to return. As much as she believes using endomycorrhizal fungi like the researchers at the University of Kansas are creating is part of the eventual solution for rewilding tall prairies—wet or dry—in the end, the most important "adjunct" could be time. "It's not that restored prairies will *never* get there," Unger said. It just might take longer than one lifetime to create one that will pass the Turing test of prairies.

For now, if you walk the spongy virgin grassland of the Wakarusa Wetlands—little changed since the last bison of eastern Kansas wandered the freakish misfit floodplain below Blue Mound—shake off the nostalgia and consider the invisible microscopic creatures doing their thing below the surface. The ancestors of those creatures were there with the bison. Descendants of other species have not yet returned, but someday might, because of the work of Unger and prairie restorationists. Those "microfauna" might be more important to future prairies than buffalo daydreams ever will be.

The sun rose hot, yellow, flammable. The woods beside the river steamed with an incandescent beauty as herons flew over, straight as laser beams, toward morning fishing spots. After the pastel light of the understory burned away, I broke camp. Twenty-seven miles remained to Kaw Point. In cooler weather, that distance would be manageable, even easy, but the previous day I'd stopped twice to keep my core temperature down when my pulse raced. I'd have to take the twenty-seven miles slow, paddling and drifting toward Kansas City and stopping often to cool off.

The Kaw makes a yawning S-loop between De Soto and Bonner Springs, at one point so narrow paddlers are tempted to portage to save distance. The river was running low. At one spot the canoe scraped bottom on a bedrock minirapids and I leaped out to avoid scratching the gelcoat. A bald eagle was perched on the handles of a grocery cart caught in a tangle of driftwood. As I drifted closer, it didn't spook. The "eagle" was a hideous clown doll, bloated and stained by the river; the kind of doll that makes children want to immediately grow up. I tried to swat it from the shopping cart and lost balance, almost falling into the river. Clammy and a little looped out from the heat, I side-pried to draw the canoe into a shallow and beached the bow in sand, tossing my chair out ahead of me. After assembling the chair on shore, I turned to grab my water jug and saw the canoe—my ride home—floating away with the current. Luckily, I was able to dive into the river and grab the stern rope.

The remainder of the morning and afternoon was a lone relay between bridges, and I fell into a good rhythm between Kansas Highway 7 and I-435. Downstream from Nelson Island at the mouth of Mill Creek, the Water One Weir completely blocks the river and requires a portage. I unloaded gear, moved the boat down, and packed everything back in, planting my remaining vitamin waters within arm's reach. The portage took twenty minutes but felt like hours. Back on the river, the current spun me around where the inlet joined the outflow running over the weir. The new canoe didn't yet fit like skin.

The lower Kaw passes several urban landmarks, but if Kansas City is a river town at all today, it is a *Missouri* river town, so even with downtown skyscrapers in sight, no actual commerce abuts the Kaw. This creates a veneer of wildness in the bottoms west of Kansas City, but it is only a veneer. Below Nelson Island, the river was filthy. Instead of willowy cottonwood fuzz, dirty suds marked the main channel. Near the Turner Bridge, I stopped to talk with a man fishing. It was his first Saturday off work in two years. Farther down I got out of the boat to rest on the last small island of the Kaw, a muddy cigar-shaped point covered in weeds: prairie clover, alfalfa, and invasive Japanese honeysuckle.

Around dinner time I reached the asphalt remains of the Kansas City Stockyards. During the twentieth century it was one of America's largest cattle centers. Extensive rail lines that serviced the stockyards are still there, but the cattle are gone. Here the Kaw sluggishly passed wastewater and chemical plants, open pipes, and straight clay culverts where unfortunate tributaries spent their final miles, diminished.

The Kaw makes its final turn to the north near Kemper Arena, where the skyline of the new Kansas City comes into sight in all its sober millennial glory. I laid my paddle down across the gunwales and slid into a calm section of reeds. Near the shore, a school of fish erupted to the surface and three carp jumped across the bow.

Along with bighead and black carp, silver carp (*Hypophthalmichthys molitrix*) are a nonnative species that is causing problems in grassland river ecosystems. Silver carp are vegetarian acrobats: they eat plankton and can trampoline their twenty to forty pounds of fish flesh high above the water when startled by boats. Introduced from eastern Asia in the 1970s as a biological control to clean algae from fish farms and wastewater treatment plants, silver carp became an invasive problem when they slipped into the Mississippi River during a flood in the 1990s. Since then, they have spread into rivers and lakes in sixteen states. Because they reproduce prolifically—a single female can lay as many as two million eggs—and consume 20 percent of their bodyweight each day in plankton and algae, silver carp put pressure on native species for food. They're also dangerous to boaters. In Illinois, a flying carp cracked the jaw of a woman who had been bowfishing.

Bowersock Dam presents a physical barrier that prevents the fish from invading the upper Kaw. In the lower river, officials from the Kansas Department of Wildlife and Parks have used electrofishing to stun schools of silver carp so they can be scooped from the river. The Missouri Department of Conservation conducts similar projects on the lower Grand. In a single weekend, teams have removed twenty-five thousand pounds of fish.

After the carp flyover, I slid back into the river for the final push toward the Missouri. My takeout at Kaw Point was where Lewis and Clark and the Corps of Discovery saw flocks of Carolina parakeets, the extinct parrot once common in Kansas prairies. Meriwether Lewis noted in his journal that the river was so clear he could see the bottom.[24] Today you'd need radar. The expedition spent three days at the confluence where they built a temporary enclosure of logs and brush.

Lewis and Clark expected to meet members of the Kaw Nation— the expedition had passed an abandoned village site earlier. Prior to their arrival in the future Kansas, the Kaw people had migrated west from the Ohio River country where they had lived with other groups known together as the Dhegiha Sioux. The exact timeline is unknown, but sometime prior to 1700 they moved west up the Missouri and split off from the Osage, who settled to the southeast. For at least a hundred and fifty years prior to Kansas statehood in 1861, the Kaw people lived in settlements between modern Kansas City and their Blue Earth Village at the mouth of the Big Blue River near Manhattan. French fur trappers named the river after the people. But the Kaw didn't use that name themselves. While the Pawnee called the entire river *Ne Miskua* (Smoky Hill), in the nineteenth century the Kaw called it *Dópik'è gaxá*, or Topeka Creek, because they gathered timpsila, the "prairie turnip"—*Pediomelum esculentum*, or in the Kaw language, *dópik'è*—near the river.

A series of treaties evicted the Kaw from their villages along the Kansas River to a reservation near Council Grove, and later to Oklahoma, where their tribal headquarters today is in Kaw City. But they never forgot their connections to Kansas, or the river. In 2002, the Kaw Nation established Allegawaho Memorial Heritage Park on 200 acres of their previous homeland near Council Grove, their first landholdings in Kansas since moving to Oklahoma.

And they never forgot about the prayer rock that had been moved to Robinson Park in Lawrence, one the few Native American religious monuments directly associated with a grassland river. Even in

the twenty-first century, most Kansans didn't know about its importance to the Kaw people and the cultural history of the river. When
Lawrence was planning a 150th anniversary celebration in 2004,
one of the organizers said in a newspaper interview that "in some
ways, the spirituality of the rock has been shifted to a very different
area. The boulder was very important to the residents of Lawrence
who brought it here—it was kind of like their Plymouth Rock."[25] He
might not have known that descendants of the Wampanoag, whose
ancestors were decimated by smallpox and sold into slavery despite
helping the Mayflower Pilgrims, hold an annual Thanksgiving "National Day of Mourning" that commemorates "the genocide of millions of Native people, the theft of Native lands and the relentless
assault on Native culture" symbolized by Plymouth Rock.[26]

According to Pauline Eads Sharp, Kaw Nation citizen and past
vice president of the Kaw Nation Cultural Committee, by 2019
sentiment toward the prayer rock had shifted. Sharp—who per-
forms Chautauqua-style historical readings about the life of her
grandmother, Lucy Tayiah Eads, the first woman chief of the Kaw
Nation—worked with Kansas-based artist David Lowenstein to develop "Between a Rock and a Hard Place," a project that brought
together research and stories of Iⁿ'zhúje 'Waxóbe to share with the
Kaw People. Together with a team of Kaw Nation citizens and Lawrence residents, they held a series of community meetings to imagine the future of the sacred boulder. When one hundred people
came to the first event, they realized that consensus was building to
redress the harm done when the rock was moved to Robinson Park.

Sharp didn't know about the sacred rock growing up. She told
me she first learned about it while helping historian Ron Parks research Kaw history for a book he was writing. She said that when
the rock was at Shunganunga Creek, the Kaw people "went there to
offer prayers to Waconda. I believe they thought of the rock and its
surroundings like a church."[27]

In December of 2020, the Kaw Nation made a formal request to
the city of Lawrence to return the prayer rock. The Lawrence City
Council issued a formal apology and passed a joint resolution to return the rock the following spring. Three years after the resolution,

and almost a century after it was pried from its ancient spot beside the river, the Kaw people reclaimed Iⁿʼzhúje ʻWaxóbe and, after removing the "founders" plaque (which was donated to a local history museum), brought the twenty-ton prayer rock home to Kaw land at Allegawaho Memorial Heritage Park.

During the peak of the commercial fishing era, Iⁿʼzhúje ʻWaxóbe sat one hundred yards from where Abe Burns skin-dove into the churn below Bowersock Dam, where I had begun my trip the day before. Now, after twenty-five years of never *finishing* the river, I approached the Missouri, a half mile ahead. Soon, my takeout point at Kaw Point, the starting line of the MR340, would appear. My heart thick with the river, I took a deep breath and pointed the bow toward the Missouri.

But it was hubris. Eleven bridges section the sky above the lower Kaw: I-435, Turner Diagonal, I-635, K32, 69 Highway, 12th Street, 7th Street, I-670, the Central Avenue Viaduct, James Street, and the Lewis and Clark Viaduct, a bridge that opened in 1907 and survived the floods of 1951 and 1993. Only one bridge, named for rivermen no less, separated me from the finish now. But the bridge was gone. A large red construction crane was lifting something from a giant pile of twisted brown metal in the river. Bulldozers and workers using arc-welding torches worked on the pile that the crane was building.

I'd never encountered anything like this on a river. With little publicity, a demolition crew had set off explosive charges at 8:30 a.m. and *dropped* the entire west section of the Lewis and Clark Viaduct into the Kansas River. The crew of a fishing boat had been trying to find their way around the demolition, but two workers standing on the crane platform looked down at us with their arms crossed and shook their heads. There was no getting around the wreckage. Portaging was impossible. I still hadn't finished the Kaw.

The pilot of the fishing boat anchored as close to the downed bridge as possible to see if the excitement attracted any fish. The crew invited me to join them, but the sun was beginning to set, and I didn't want to get caught on the river at night. The final quarter mile of the Kaw would have to wait, perhaps until the start of the MR340.

Rivers encode worldviews in their watersheds. They devour the familiar. I wasn't sure that I knew Burns's River—Tom Burns's or Abe Burns's. How different was it from the days when Jake Washington taught himself to fish with hoop nets and commercial fisherman plied the six miles below Bowersock Dam? For years now, the section has been so polluted with carcinogenic polychlorinated biphenyls (PCBs) that KDHE warns people to eat no more than one serving per week of bottom-feeding fish such as flatheads. How different was it from when the Kaw people lived near the prayer rock? But maybe none of that mattered. Most people who paddle Upper Missouri Breaks National Monument come from outside Montana. Despite a dozen well-developed access points and startling potential, few paddle the Grand, at least today. The Kansas River, however, is a beloved *home* river to thousands who return to it each year, "through-paddling" all 173 miles in sections, turning friends on to favorite sandbars, nodding at eagles they know on a first-name basis, marinating in the essence of the longest grassland river in America.

A delicious north wind began to blow, the first relief from the heat in two days. It sailed me back to Kemper. Christina waved from the top of the levee. After unloading my gear on the floating dock, I lifted the canoe from the water, gentling it down onto the metal grating. I'd reached Missouri the state, rather than Missouri the river, but it was a taste. My time on the big river would come.

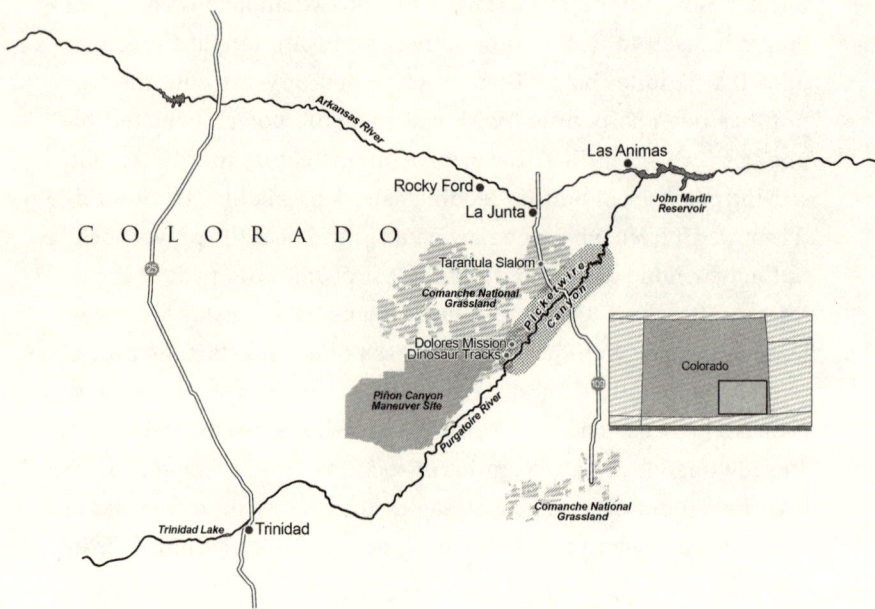

MAP 5 The Purgatoire River

CHAPTER 5

PURGATOIRE

As it flows into the plains of southeastern Colorado from the foothills of the Sangre de Cristo mountain range, the Purgatoire River is haunted by footsteps: of migrating tarantulas, of ancient canyon people and their rock art scattered across silent boulder glens, of long-departed Hispano settlers whose culture can still be traced from California to the southern plains, and of monsters. Near the Purgatoire's confluence with the Arkansas River, Picketwire Canyon boasts North America's longest dinosaur trackway, a film of bedrock where lumbering apatosauruses pressed their bird-like feet into the wet mud of a Jurassic lake, creating a fossilized Polaroid from an era when giant ferns and sequoias blanketed the future American interior. Remoteness and danger guard these fragile footsteps. Hikers still die in the harsh microclimate of the canyon—a stunning yet devastating landscape that has run people off for centuries. And the treacherous beauty was backdrop for a different kind of struggle, when the river and surrounding Comanche National Grassland became ground zero in a fight to save private ranches and public prairies from a proposed expansion of Piñon Canyon Maneuver Site, a large-scale military training facility. The battle forged an unlikely alliance between ranchers, dinosaur hunters, and environmentalists.

After paddling the Grand and Kansas Rivers, I wanted to backtrack from the well-watered realm of tallgrass prairies to learn about shortgrass prairie rivers in the Great Plains. Two such areas in Colorado stood out. A singular stream, the Purgatoire has a vivid natural and cultural history. For most of its compass, the river is too shallow to propel a loaded canoe at normal flows, but ancient footpaths in its canyonlands make perfect avenues to explore the idiosyncratic river. Farther north, the South Platte carries meltwater from the Continental Divide through the heart of downtown Denver and out onto the high plains of Colorado and Nebraska, where drought and climate change are intensifying battles over water rights in the Great Plains.

Colorado, of course, evokes images of mountains, not prairies. Its highest peak—the 14,400-foot summit of Mount Elbert—is the second-highest point of land in the United States outside of Alaska (California's 14,505-foot Mount Whitney is the highest peak in the lower forty-eight). Colorado has the highest average elevation (6,800 feet) of any state and more 14,000-foot peaks (fifty-eight) than the rest of the states combined. However, a significant chunk of the Centennial State is in the Great Plains. Grasslands covered 40 percent of future Colorado in pre-European settlement times.

Video game development companies, trail runners, snowboarders, legalized pot dispensaries, and a quasi-religious devotion to locally made beer are twenty-first century newcomers to the state, but nostalgia for the days of gold mining, cattle rustling, and the "Wild West" remains. Tourists still flock to Victorian-era downtown districts in tin-type towns like Victor and St. Elmo, ride the Durango and Silverton Narrow Gauge Railroad, and eat elk steaks and Rocky Mountain oysters at the Buckhorn Exchange, a restaurant that has operated from the same Denver location since 1893. Southeastern Colorado, however, is different. It was not part of the 1803 Louisiana Purchase that set the stage for Colorado's Western identity. The region was influenced more by an earlier Hispano era of European settlement, a history that resonates to this day.

The 196-mile-long Purgatoire River arises in the Culebra Range of the Sangre de Cristo Mountains, descending rugged slopes before entering one of the largest regions of privately owned forest in Colorado. Completed in 1958 to prevent downstream flooding, the discharge controls at Trinidad Lake have mellowed seasonal flows driven by spring snowmelt and summer monsoons. East of Interstate 25, the river switches to another lens, zooming into a region of plateaus, mesas, and deep gorges before reaching the full-on Great Plains at Comanche National Grassland, home to the magnificent 16,000-acre Picketwire Canyon. There, for a dozen miles, the Purgatoire is as wild as any grassland river in America.

Prior to the arrival of Europeans, the Jicarilla Apache, Comanche, Arapahoe, Cheyenne, and Pawnee frequented the Purgatoire River valley. The Cheyenne called it *Hōtoǎnǎ'ohe*, *"Difficult* River." In the sixteenth century, when an entrada of Spanish soldiers vanished along its main fork, Catholic priests in Santa Fe christened it *El Río de Las Ánimas Perdidas en Purgatorio*. French fur trappers later shortened it to *Purgatoire*. American settlers transliterated it *"Picketwire,"* easier for Americans to say, but meaningless. Despite the French suffix, today everybody pronounces it "Purga*tory*." I would soon learn why the name's original Spanish meaning—*River of Lost Souls*—might be more appropriate.

On the lone paved road between La Junta and Picketwire Canyon, Colorado Highway 109 is nicknamed the "tarantula slalom" because of the cars that swerve for spiders during their peak migration in early autumn. Not far outside of town, dryland corn-stubble fields dissolve into a tea of orange and yellow grasses, a steeping emptiness that intensifies beyond the entrance to the 440,000-acre Comanche National Grassland. On early maps, the word *"wasteland"* was sometimes lettered across the region in the shape of an ironic smile.

The three cars of our caravan plumed dust clouds that cleared whenever the road thinned to jeep ruts. My friends and I were trying to reach camp before nightfall—the sun set at 4:30 p.m. in early December. A potential winter storm was in the forecast. Wind-sculpted snowdrifts in north-facing ditches remained from a recent arctic pulse. I hoped any new snowfall would hold off long enough for the six of us to hike Withers Canyon Trail down into the wildest part of the Purgatoire River valley. Next to the chalky road, a thin antelope grazed the slope of a low mesa. The hazy silhouette of Pikes Peak, one hundred miles distant, loomed on the horizon like a pale blue planet.

The river's longest run through public lands is in the Picketwire Canyon section of Comanche National Grassland, but it flows intermittently. Below the grassland, there is sometimes enough water to prod a canoe or kayak the final ten miles to the confluence with the Arkansas River, but access is difficult. I had twice spectacularly failed in attempts to paddle between the only two points of public land along the route—Purgatoire River State Wildlife Area and a city-owned access point in Las Animas. On the Colorado plains, accessing the river via public land is a requirement. Compared to other western states, Colorado river-access laws are ambiguous when it comes to crossing private lands. According to the Colorado Whitewater Association, boaters have the right to paddle most of Colorado's rivers if they don't trespass on private property to get on and off the river.[1] But on shortgrass prairie rivers like the South Platte and the Arkansas, there have been disputes between paddlers and property owners. Colorado has no codified law that recognizes a right to walk up to the high-water mark of rivers. Paddler rights groups cite the Colorado Attorney General's 1983 opinion that permits downstream passage on all waters of the state if there is no contact with the beds or banks of the stream.[2] With more than a dozen landowners along the lower Purgatoire, obtaining permission was impractical, so paddling the entire section without leaving the boat was the only way.

On one trip, the river tottered near flood stage. After carrying the canoe for one hour road to river, eddies frothed in the swirl, the full torrent of the river caged in a channel not twelve feet across. Possible drowned barbed wire was more worrying than the cottonwood strainers spaced at regular intervals downstream. I made it across the state wildlife area to the edge of private property before the sick green current, jaundiced with sediment, nearly flipped my boat and tossed me into the churn, ending the trip. It was one of the most harrowing thirteen minutes of flatwater in my river years. On another visit, I alternated floating and dragging the canoe across the same section and again turned back rather than roping the boat through the prairie on foot. Nobody, I found out, paddles the lower Purgatoire.

Ready to learn from others, this time we had come to backpack— not paddle—the valley, on foot trails that people had used to traverse the riverscape for centuries.

Comanche National Grassland is the fourth largest of America's national grasslands. The national grasslands were not created to enshrine unspoiled ecosystems, instead they have a backward sounding history. Their origins trace back to New Deal government land buybacks aimed at mitigating the results of an environmental catastrophe brought on by drought and ill-fated agricultural practices more than a century ago. Beginning with the Homestead Act of 1862 and accelerating in the late nineteenth and early twentieth centuries, agricultural entrepreneurism, new technologies, and blind faith lured Americans west to the erstwhile "Great American Desert" to try their hands at farming cheap or free land. American citizens (and European immigrants intent on citizenship) could claim 160-acre "homesteads" if they remained on the land for at least five years and *improved* it. Plowing native grasslands qualified as one type of improvement.

The prospect of free land and a new start has always been an American bugle call, but it was an especially potent aphrodisiac near the end of the western expansion period. Many Great Plains homesteaders had little or no farming experience. Over a seventy-year period, they plowed millions of acres of grasslands, causing the population of plains states to surge from less than a million in the 1870s to five million by the mid-1930s.

For good reasons, row crop agriculture came late to the western Great Plains. Mountains along the Continental Divide sip moisture from Pacific airmasses, creating a rain shadow east of the Rockies where shortgrass prairies thrive. Before the invention of center-pivot irrigation made it economically feasible to mine water from the Ogallala Aquifer, crops like corn and wheat required twenty inches of rainfall per year to produce a salable yield. On average, in one of every three years, that minimum threshold is not reached. Droughts are the real old-timers of the region, as indigenous to the landscape as bison and prairie dogs. A tree-ring chronology from western Nebraska showed that droughts of at least five years happen on average every twenty-four years.[3] One of the deepest and harshest droughts spanned a thirty-eight-year period between AD 1276 and AD 1313.

In the 1930s, agricultural aspiration exceeded precipitation by a disastrous ratio. A deep drought set in and lasted for a decade accompanied by extreme, unprecedented heat. Record high temperatures from the summers of 1934, 1936, and 1937 have gone unbroken even in the era of man-made climate change. The drought withered crops and native vegetation, exposing millions of acres of pulverized soils where prairies had been plowed. Overgrazing of the remaining uncultivated rangeland worsened the situation. The aftermath of the stock market crash of 1929 and the collapse of commodity prices following World War I chainsawed heartland farming economies and set the stage for the Dust Bowl, when winds blew porous soils in great storms that reached New York City, Boston, and the decks of ships sailing in the Atlantic Ocean. Blizzards of snow and dust called "snusters" killed people in their homes. Hunger and bankruptcy bludgeoned farmers in a vertical strip from Texas and the Oklahoma panhandle north through eastern Colorado and western Kansas. By the end of the 1930s, almost one trillion tons of soil and most of the unplowed shortgrass prairie had vanished.

Soon after he was elected in 1932, President Franklin Roosevelt began enacting policies to bail out and resettle destitute farmers and keep the soil of the Great Plains in the Great Plains. The National Resources Board, created by a 1934 executive order, recommended an astonishing government buyback of some seventy-five million acres doled out during homesteading. Congress authorized the National Industrial Recovery Act of 1933, the Emergency Relief Appropriations Act of 1935, and the Bankhead-Jones Farm Tenant Act of 1937 to provide funding for the purchase and rehabilitation of distressed lands, and eventually eleven million acres were acquired—about one-sixth of the original recommendation. Management of the new federal lands was split according to the region between the Bureau of Land Management (BLM) and the US Forest Service. In 1960, four million acres were cordoned off from US Forest Service lands to create the national grassland system.

Twenty national grasslands exist today. Crooked River National Grassland in Oregon, Butte Valley National Grassland in northern California, and Curley National Grassland in Idaho are in western states, but the remaining preserves—Cedar River, Sheyenne, and Little Missouri National Grasslands in North Dakota, Thunder Basin National Grassland in Wyoming, Grand River and Fort Pierre National Grasslands in South Dakota, Little Missouri and Oglala National Grassland in Nebraska, Cimarron National Grassland in Kansas, McClellan Creek, Lyndon B. Johnson, Rita Blanca, and Caddo National Grasslands in Texas, Kiowa National Grassland in New Mexico, Black Kettle National Grassland in Oklahoma, and Comanche and Pawnee National Grasslands in Colorado—are all in the Great Plains. It may sound strange, but the national grasslands are managed by the US Forest Service. Like the national forests, the grasslands are managed for "many uses." In early years, that meant revegetation and soil stabilization, later expanding to grassland conservation, wildlife management, recovery of threatened and endangered species, cattle grazing, mineral, oil, and gas mining, hunting, fishing, and outdoor recreation.

Even more than in other Dust Bowl–ravished regions, epic drought punishes the high plains of southeastern Colorado with clock-like regularity. Not far from the northernmost Chihuahuan Desert, the flora and fauna sometimes take drastic survival measures. For example, the striped bark scorpion can "*aestivate*," entering a rigid state of torpor during summer droughts until rainfall and cooler weather return. The average precipitation in La Junta, on the edge of the northern unit, is less than twelve inches per year.

David Augustine, a research ecologist for the US Department of Agriculture, has studied how the extreme environment affects both endangered and invasive species at Comanche National Grassland.

Augustine has a wide-ranging portfolio of publications on short-grass prairie management, covering topics such as cattle grazing, prescribed burning, and the ways that prairie dogs affect grassland quality and the habitat of plains bird species like the federally threatened lesser prairie-chicken. He told me, "Most of the pastures at Comanche National Grassland, and all the national grasslands for that matter, were homesteaded, which meant they had a major human footprint. They're fragmented, with federal parcels adjacent to private land in a checkerboard pattern of ownership. Because the purpose of the grasslands originally was soil stabilization, for most of their history, they have been managed as extra pastures for adjacent ranches."[4]

Some of the preserves, however, harbor rare plant and animal communities. Comanche National Grassland was, until recently, home to 25 percent of the lesser prairie-chickens in Colorado, and possibly up to 5% of their total worldwide population. The grassland is comprised of two units: the Timpas unit in Otero County and the Carrizo unit in Las Animas and Baca counties. Augustine said the sand-sage prairie habitat that lesser prairie-chickens need is concentrated along the Cimarron River in the Carrizo unit, rather than along the Purgatoire farther north in the Timpas unit. Following a series of harsh winters and blizzards, by 2016 the population of lesser prairie-chickens at Comanche National Grassland had dropped to fewer than fifty birds. Since then, Colorado Parks and Wildlife has been reintroducing birds trapped in Kansas in hopes of stabilizing the species. In 2022 the lesser prairie-chicken was declared threatened in Colorado under the Endangered Species Act.

I wasn't concerned about our odds of seeing lesser prairie-chickens along the Purgatoire. With a winter front coming in and blizzards on my mind, we wagoned our tents around a metal hearth beside Withers Canyon Trailhead and tried to stay awake long enough for the last embers of the fire to coal over. During the night, boreal air shivered me awake and I rolled over to check the pull thermometer on my backpack: 22 degrees Fahrenheit. I tried to warm up by stepping out of the tent and walking into a thin grove of piñon pines, hoping to get a glimpse of the river winding through the gorge below. Yet the frigid moonlight seeping into fissures and cracks in the rocks did not penetrate the black maw of the canyon. The lonely throbbing of a strobe light on a distant microwave tower was the only reminder of the made world. It made me think of the lyrics of the Jason Molina song "Farewell Transmission." Before he died young, Molina wrote dark meditative tunes, apt communion for all-night cross-country drives or restless souls struggling to sleep near the edge of haunted canyons.

The next morning, we scrutinized each other's water stores. Hydration is crucial in Picketwire Canyon, even in December. I divvied up a gallon of water among hydro flasks colored like 1970s bean bags. Our plan was to hike six miles to a knee-deep cold-water crossing of the river to explore the dinosaur track site. On our return, we would search off-trail boulder fields for petroglyphs.

It seemed strange to explore a river on foot. With a snowstorm possible, it was good to be with my five companions, who were all Kansas River guides and knew what they were doing in remote wilderness settings. Packs and water hoisted, we paused at the trailhead to read a US Forest Service sign:

Caution, hiking into the canyon is OPTIONAL. Hiking out of the canyon is MANDATORY. Self-rescue may be your only option. If you encounter heat related troubles, return to the river and, with clothes on, completely immerse yourself, rest in the shade, drink river water, and wait until early evening or night to exit. This may save your life.

A bit wordy perhaps, but the ominous welcome was no bluff. The afternoon high temperature wouldn't top 45 degrees Fahrenheit, yet Picketwire Canyon had already begun sipping moisture from our soft tissues as we navigated switchbacks 250 feet down to the can-

yon floor. We had crossed an invisible boundary into a harsh microclimate that didn't yet feel harsh, and therein lay its danger. In 2017,
two hikers died from heat-related injuries on the trail.

The canyon floor felt like the bottom of the world, the stillness
of the place jarring. The soft sunrise revealed frozen spiderwebs
anchored between dead spikes of yucca. The black cold of the previous night retreated into long shadows cast by cliff strata. Piñon
pines and juniper trees dotted fields of straw-colored buffalo grass.
The transition zone from Great Plains to canyonlands was visible in
the distance, beyond a line of orange willow hugging the riverbank
in a mile-wide floodplain surrounded by low bluffs on either side.

Cactus land-mined the prairie. Cholla with withered yellow
flowers filled every niche. Forests of fruit-heavy prickly pear and
needle-latticed clumps of red cactus grew in rockier patches. Up
ahead, someone asked how to pronounce "cholla" and I recalled a
rhyme from geography field camp: *Nice to know ya', cholla.* Cholla
are the extreme sports of southwestern and Sonoran cacti. *Cylindro-
puntia imbricata,* the tree cholla, is a *flying* cactus that can break off
and hitch a tortuous ride on passing haunches. The tallest plants in
the floodplain except willows and arthritic cottonwoods, the fuzzy
succulents were ubiquitous, like daggered dandelions conquering
a feral lawn.

David Sain stepped off the trail to check his dog Cassidy's paws
for prickly pear tines after she raced a jackrabbit around the side of
a juniper. A river guide and architect of space-age homes that other
designers had taken to calling "Midwestern masterpieces," he once
prickled when a waitress in a small New Mexico town looked his
outfit up and down and said, *"men don't wear scarves around here."*
David had come to share the canyon with friends, see the dinosaur
tracks, and hike the mostly flat riverside trail—in that order of importance. More accustomed to solo treks in the Rockies, the twelve-
mile hike was light calisthenics for him. It would have been easy
for me too, but I'd had surgery the month before and was nervous
that my meniscus-trimmed knee would slow the others. Fingering
the lid of my water bottle, I heard what I thought was a muffled cry

out in the sage. David thought he heard it too. The canyon's qui-
etude was playing tricks on us, the silence almost incantatory. The
sound was likely nothing more than an auditory mirage, certainly
not the murmuring souls of lost Spanish conquistadores. We finished drinking our water and moved on down the trail.

Three miles later, we still hadn't glimpsed the actual river, although it was only a few hundred feet out in the grasses. When
the path climbed higher into the bluffs, we ascended a hill of loose
boulder shavings on a rind of trail that led to the top of a cliff. The
sweep of the flat plain of the Purgatoire came into full view, and we
saw the sparkling river itself, a hundred feet wide, snaking ruddy
through eddy pools and gravel bars below. A willow grove and an
old Russian olive—the noxious exotic imported from Asia in the
1890s and fought today as a pest—grew in the gravel, along with at
least one tamarisk shrub, its summer blooms dissolved into smoky
lavender gruel.

Controlling Russian olive and tamarisk is a conservation priority at many of the national grasslands. When David Augustine, the
research ecologist, started working at Comanche, their primary focus was eradicating tamarisk, the most urgent ecological concern
along the river. He explained, "Tamarisk affects cottonwood galleries, and when I started, the river corridor was a continuous band
of tamarisk."

Due to their natural drought resistance, native grasses and forbs
in the uplands—even in restored areas—typically fare well against
invasives. Grassland river bottoms, however, share the Achilles'
heel of arid rivers across the American Southwest. *Tamarix ra-
mosissima*, a squat shrub introduced as an ornamental in the 1880s,
has spread like crabgrass along drainages. Tamarisk can decimate
riverscapes: its deep taproots draw down stream flows, particularly
during droughts, and each shrub creates a toxic saline perimeter
around it that puckers native plants. Tamarisk clot streams, widen
narrow river runs, and diminish habitat for willow-nesting birds.
In wildland fires, tamarisk ignites like a dried-out Christmas tree
thrown onto a Valentine's Day bonfire.

Even methods for combating the plant have proved contentious. Herbicides, sawyers, and goats have been employed. Biologists in southwestern states have imported tamarisk-chomping salt cedar leaf beetles from Kazakhstan as a cheaper and more ruthless biological control agent, despite obvious concerns that, like tamarisk itself, nonnative beetles could impact riparian ecosystems in unforeseen ways. In Utah, the beetles do such a thorough job of consuming tamarisk leaves that second-wave invaders like Russian olive and perennial pepperweed can swiftly occupy the vacated niches. Everybody agrees that the ultimate goal is to replace tamarisk with native species. Since 2005, a coalition of public agencies, the nonprofit Tamarisk Coalition, and private landowners have battled tamarisk along the Purgatoire using the "cut-stump" method. This approach involves scissoring off the shrubs at ground level and treating the stumps with herbicide to discourage metastases.[5]

Along this stretch, the native willows outnumbered tamarisk and Russian olive combined, suggesting that the coalition's efforts seemed to be working. The river reminded me of the upper Missouri, flowing at ground level in the plain, connected to its grassland surroundings, with no road, or powerline, or fence in sight. But its most untamed quality wasn't visible from the trail.

In its lower reaches, the Purgatoire is a shortgrass prairie river. Shortgrass rivers differ from tallgrass prairie rivers in several ways. Born of mountain runoff, shortgrass rivers rollercoaster down into the plains, ratcheting up stream velocities. In contrast, low-gradient tallgrass rivers are mall walkers. Shortgrass river flows boom and bust with droughts, groundwater pumping, diversion for irrigation, spring snowmelt, and summer monsoons.

This inherent dynamism affects fish found in shortgrass rivers. Like the biodiversity of grasses and forbs in unplowed prairies, the biodiversity of fisheries in a river can be graded by the presence of native species, the absence of nonnative species, and the strong numerical dominance of the former. By this rubric, the Purgatoire is the wildest grassland river in America. A 2017 study that resampled sections of the river inventoried a quarter century earlier found the river's fishery virtually unchanged.[6] A remarkable 99.6 percent of individual fish captured belonged to twelve native species: red shiner, fathead minnow, green sunfish, sand shiner, flathead chub, longnose dace, white sucker, central stoneroller, northern plains killifish, black bullhead, channel catfish, and suckermouth minnow. Only three nonnative species were detected in the river: the creek chub, common carp, and largemouth bass.

When largemouth bass encounter plucky fish like the fathead minnow, the larger fish think: appetizers! But the Purgatoire stonewalls nonnative bullies that venture upstream from John Martin Reservoir. Bob Bramblett, whose research from the early 1990s established a baseline for later studies, told me, "Prairie fish as a group are very tolerant of extremes. Prairie rivers are warm in the summer, cold in the winter. They're flashy. They flood. They dry up. There's high salinity and sometimes mud. As far as fish are concerned, these rivers are not great habitat. The fish in those rivers reflect the harshness of the environment."[7]

Bramblett found largemouth bass at only one sampling site. He said, "It's not like a trout stream where you can dump brown trout and rainbow trout on top of cutthroat, and they find a ready home. If you release a trout in the Purgatoire, it's going to die almost immediately. It's the same for sports fish like bass that swim up from downstream reservoirs. Grassland fishes possess local adaptations that help them deal with challenges posed by turbidity, intermittent flows, scalding summer temperatures, and low dissolved oxygen levels. Turbidity is especially important. Muddy waters cause problems for sight-feeding sports fish."

The Purgatoire's undulating flow patterns have been less affected by the era of reservoir building than other shortgrass prairie rivers, in part because thunderstorms in the immediate vicinity of the canyon contribute substantially to seasonal flooding. The resilience and composition of the fish assemblages in the Purgatoire River underscore its status as one of the wildest grassland rivers in America. When I asked Bramblett what was most important to ensure this continues, he said, "The harshness of the canyon does the work. We just need to stay out of its way."

There were reminders of a recent summer deluge on a turn in the far distance, where the river cut deeper into the mushy dirt-cobble and a clear line of debris was piled up on shelf rock. Beyond that, the yellow prairie grew all the way down to the willow that ribboned west until a long mesa blocked the view. At the bottom of the cliff, the trail crossed a gulch beside the ruins of Dolores Mission, the shell of a collapsed church that was part of a late nineteenth-century Hispano settlement near the riverbank.

For at least five thousand years, indigenous peoples crisscrossed southeastern Colorado between the Pueblo canyonlands of New Mexico and the buffalo hunting grounds of the western plains. Perhaps because of the lack of dependable rainfall, they left scant evidence of long-term settlement. The Spanish were the first Europeans to reach southeastern Colorado and the first who failed to colonize it. The Purgatoire and Arkansas Rivers were at a nexus between New Spain to the southwest, the French who made occasional forays from the northeast, and the Comanche and Jicarilla Apache to the east. Santa Fe's colonial governor—the brutal Juan de Oñate, who married the great-granddaughter of an Aztec emperor—eyed all northern New Mexico and southeast Colorado for Spanish plunder. In 1599, Oñate's forces massacred the Acoma Pueblo, killing eight hundred people. He ordered his troops to amputate the feet of all surviving men over the age of twenty-five. In 1997, somebody sawed the bronze foot off an Oñate statue in Alcalde, New Mexico, in the middle of the night. Years of Pueblo revolts, the rise of the Comanche after they perfected the art of war on horseback, and the Catholic Church's disapproval of territorial Santa Fe's reluctance to Christianize Native Americans kept Spain from settling the lands along the Purgatoire.

In 1821, Mexican independence changed the dynamics of its northern provinces in New Mexico and Colorado. Forbidden un-

der Spanish rule, legalized foreign trade opened the door to an influx of French-Canadian fur trappers and American traders traveling through the Purgatoire River valley on the Santa Fe Trail. Thirty
years later, America would establish a new southern border and
take all of New Mexico and Colorado after the Mexican-American
War; but between 1840 and 1847, the territorial governor of New
Mexico issued twenty-three land grants under Mexican law designed to create a buffer against the Comanche and the increasing
onslaught of Americans. In contrast to homesteading grants that
the United States would later employ to crowdsource its Manifest
Destiny 160 acres at a time, just one of the Colorado grants—the
1843 Vigil and St. Vrain or "Las Animas" Grant—encompassed almost the entire Purgatoire River valley and totaled more than four
million acres.

After Mexico ceded New Mexico, California, Arizona, and parts
of Utah and southern Colorado to the United States, the validity
of the Las Animas Grant was immediately challenged. Congress
told landowners to bring Spanish and Mexican grants before the
Surveyor General's Office for validation, a process that took decades. To shore up support for their claims, grant holders recruited
Spanish-speaking settlers, known as *Hispanos*—former Mexican citizens born in New Mexico, Arizona, and Texas—to establish settlements and communities in the Purgatoire River valley. Between the
1850s and 1870s, hundreds of New Mexican pioneers came to future southeastern Colorado and built adobe houses and towns with
plazas. Many of the Hispano settlements didn't last long due to the
harsh environment and Native American attacks. Towns like Trinidad, Aguilar, and Walsenburg survived and trace their roots back to
this original Hispano migration, loosely mapped to the prevalence
of green chile stew on modern restaurant menus. (This endemic
dish of New Mexico and Colorado is like a thick creole soup that's
been stung by a scorpion. It typically includes potatoes, onions, gar-
lic, oregano, pork, and fire-roasted green chile peppers.)

Hispano emigration began to wane about the time that Colorado
entered the union in 1875. By then, the Purgatoire valley was majority Hispano, with pockets that remained so well into the twentieth

century. The region was devastated by the blizzard of 1886 and a drought that burst the bubble for both Hispano and Anglo settlers and began a period of population contraction.

As we walked up the gulch a short way to the mission, the ruins were starched testimony to hard times in a difficult land, picked clean by the scorching decades that followed. Only the broken walls of the clay and stone church remained, with two small crosses propped against a wall. Gravestones of a dozen Hispano settlers stood by in the yard. A flock of mountain bluebirds landed on the ruins of the church like Christmas lights. The site blends so seamlessly into the canyon's piñon-studded foothills that I wondered if the founders had intentionally camouflaged it, either from Comanche attack, or so the brutal heat and drought might spare what it couldn't find. After the blizzard of 1886, the small town deeded the mission and cemetery to the Catholic Church, but it eventually fell into ruins. In 2007, the church donated it to the Forest Service to preserve the direct connection to the Hispano settlement period.

Outside the mission, someone noticed faint tracks the size of raindrops that led from one of the graves to a big cholla. David motioned me over to check out a hole lined with pillowy silk in the dirt. We gathered around to examine the dried-up remains of a tarantula. It wasn't alone. After we knew what to look for, we started finding more scooped-out domiciles along the perimeter of the old mission. Something, it seemed, was still pioneering the valley of the Purgatoire River.

Aphonopelma hentzi, the Texas brown tarantula, is the only tarantula species native to the western shortgrass prairies, and arachnophobes, get this: each September the hairy giants *migrate*. By moonlight . . . by the thousands . . . . Tarantulas are part of the lore of southeastern Colorado, especially near La Junta in Comanche National Grassland.

Arachnid researcher Hank Guarisco has spent months living along the Purgatoire while studying its spiders. He told me the ta-

FIGURE 5.1 Texas brown tarantula (*Aphonopelma hentzi*). Photo courtesy of Hank Guarisco.

rantula migration isn't like other great animal migrations, say of passerines or the Porcupine caribou herd. Males emerge from their dens in September to look for mates. Because the entire population does this at the same time, people mistake annual arachnid movements for true migration. But to the casual observer, the effect is the same.

Once when Guarisco was driving near a rock quarry at sundown during the second week of September, he noticed a "weird motion at the edge of the road." Guarisco said, "I pulled over because I thought it might be a tarantula. It was very strange. When I got out of my car, I found a tarantula hawk dragging a tarantula across the road."[8]

Only an arachnid researcher—and perhaps only Hank Guarisco—would equate *"weird motion"* with "tarantula" while driving at highway speeds. Tarantula hawks are shiny black wasps with dark red wings (like the colors of a scarlet tanager inverted; the colorful birds have red bodies with black wings). The state insect of New Mexico, adult tarantula hawks feed on nectar most of the year, but when females get ready to lay their eggs, they become spider assassins. Tarantulas are defenseless to the wasp's venom; their only hope for survival is to run. The venom paralyzes the spider but doesn't kill it outright. After stinging it, the wasp flies the spider back to the den for a macabre sort of meal prep. After arranging the spider, she lays her eggs on it so that when the young emerge, their first meal is ready for them. By that time, the mother has long since gone back to sipping nectar.

People describe the sting of the tarantula hawk as one of the most painful in nature. The entomologist Justin Schmidt has said that some people, when stung, lay down and scream.[9] He has described the feeling as instantaneous, electrifying, and debilitating, though brief. The pain lasts about five minutes and does no permanent damage.

Like others who've spent time alone on the Purgatoire, the canyon played tricks on Guarisco. "I stayed four days down there once in June when it was really, really hot. I was staying in a Forest Service bunkhouse with a swamp cooler. I slept during the day, during the heat, and came out in the evening to do my work. After four days of that, being alert, watching everything, hiking in the evening, when I drove out and came back to La Junta, it was mind blowing. A bunch of school buses were parked around a water tower. To me, they looked like cattle around a stock tank. My mind was changed by being in the canyon so long." The canyon did feel isolated. When a hawk disappeared over the rim of the gorge, it seemed to fly off the face of the earth.

After another twenty minutes on the trail, we found what looked like a gigantic shoulder blade on the side of a small hill. Closer inspection showed it was just that—the eight-foot-long shoulder blade of an apatosaurus that had been pried from a nearby rock quarry or, at least, a cast replica of it. A big Jurassic joke, it seemed, and I was tired enough to fall for it. Anybody who played with dinosaur toys as a kid is familiar with the 40,000-pound apatosaurus. Paleo-classifiers called the species brontosaurus before declaring it a separate genus in 2015. The corny art installation was the largest man-made artifact of the last five miles. It gave me a second wind,

FIGURE 5.2 Cassidy standing in dinosaur tracks at Picketwire Canyon, Comanche National Grassland. Photo courtesy of David Sain.

and soon the trail sickled toward the river around a final curve. David pointed toward an outdoor bathroom and Forest Service signs. We had reached the fast-flowing Purgatoire River and a matrix of bedrock that, we soon found, contained the souvenirs of real dinosaurs.

In the late autumn of 1935, fourteen-year-old Betty Jo Riddenoure let her science teacher in on a secret. She and her family knew of a strange set of depressions in a quarter-mile scour of bedrock not far from where she lived. The depressions were often covered with mud and debris, especially when the Purgatoire River, adjacent to the site, overflowed its banks during seasonal floods. They called the place Elephant Crossing, and Riddenoure often fantasized about what strange creatures could have made the marks in the stone. When her science teacher organized a field trip, a reporter for the local newspaper came along and wrote a story about the mysterious impressions along the river.[10]

That winter, John Stewart MacClary, a Pueblo-based outdoors writer who was wheelchair bound after an accident had paralyzed him in 1929, read the story of the depressions and organized a larger expedition. Piloting a 1920s-era jalopy down into the remote canyon, a group of MacClary's friends found the crossing, took hundreds of photographs, and returned to Pueblo with stories of monsters. Armed with the photos and field notes from his friends, MacClary tried to stir up national interest in what he believed were the fossilized footprints of actual dinosaurs taking a walk together. He eventually wrote an article about the site for the first weekly edition of *Life Magazine*.

By the time the world learned about Betty Jo Riddenoure's secret beside the Purgatoire, Colorado was already famous for dinosaurs. The state was a principal battlefield in the "bone wars" of the nineteenth century, an era when dinosaur hunters scoured the canyons and arroyos of the American West. The likes of Edward Drinker Cope from the Academy on Natural Sciences in Philadelphia and Othniel Charles Marsh from Yale tried to claim Jurassic plunder from the greatest fossil troves ever discovered. First excavated in 1877, "Dinosaur Ridge" in Morrison, Colorado, was the most famous dinosaur dig on earth at the time.

Gargantuan skeletons of dinosaurs—stegosaurus, triceratops, tyrannosaurus rex, megalosaurus, and others—were centerpieces of Gilded Age museum collections, but in the 1930s and 1940s, paleontologists had not yet turned their attention to dinosaur tracks. Roland Thaxter Bird, a fossil sleuth from the American Museum of Natural History in New York, visited the Purgatoire after he read MacClary's article and declared the tracks were made by brontosauruses, the first brontosaurus tracks ever documented. Bird was on a whirlwind fossil-finding trip that summer and, just a few days after leaving southeast Colorado, discovered another set of brontosaurus tracks on the Paluxy River in Texas. These tracks weren't at the bottom of an inhospitable canyon. In large part because the latter tracks were easy walking distance from a road, Bird focused his work on Texas, and interest in the remote Colorado site waned. For the next fifty years, few people studied or even visited the remarkable tracks along the Purgatoire River.

In 1980, not long after moving from Wales to teach at the University of Colorado, Denver, Martin Lockley was told by a colleague about the brontosaurus tracks in the southeastern corner of the state. Lockley, a geologist and paleontologist who didn't know much about dinosaur tracks, unearthed a few obscure papers that had been published about the site. After corresponding with Betty Jo Riddenoure, who by then was in her sixties and living in Arizona, Lockley gathered maps and visited Picketwire Canyon for the first time in 1982. When he reached the track site, he couldn't believe that such a detailed record of dinosaur behavior had gone unheralded in a state famous for dinosaur discoveries. During dozens of subsequent visits, Lockley created a meticulous spatiotemporal analysis of five sets of parallel tracks. Although years earlier Bird had theorized that fossil tracks held clues to the social behavior of dinosaurs, Lockley intended to prove it.

When I spoke to him about his early days at the site, he told me, "We knew the trackways on the Purgatoire were laid out in parallel, but after we made careful measurements, we found they had very regular spacing, even to the point that if one veered off to the right then they all veered off to the right, then back to the left. The pattern repeated itself and was very organic."[11]

Lockley and his fellow researchers found geological evidence of a lakeshore and theorized that a group of dinosaurs, perhaps a family, were walking parallel to its banks, like people taking a walk on the beach. "This is a very common pattern," he said. "Along oceans or large lakes, there is a tendency for big animals to walk parallel to the shore. The evidence gives the impression that they were walking on a broad front, shoulder-to-shoulder, rather than one going

FIGURE 5.3 Purgatoire River in Picketwire Canyon, Comanche National Grassland. Photo courtesy of David Sain.

first, then another following, crisscrossing their tracks. This was gregarious behavior."

Lockley's seminal paper, "North America's Largest Dinosaur Trackway Site," was published on the fiftieth anniversary of Roland Thaxter Bird's one-and-only visit to southeastern Colorado.[12] The paper made Lockley an instant celebrity in the paleontology world. The publication's timing was auspicious. Paleontology, as a field, had moved on from merely collecting the bones of dinosaurs to analyzing their behavior, their social patterns, even their personalities and moods. Lockley established himself as an expert in the new field and traveled the world to study the fossilized tracks of different species of dinosaurs, extinct mammals, and early humans. He documented a site with fossilized claw marks that he and his colleagues believed was evidence of dinosaur sex—or at least foreplay—because the sandstone scrape marks were like those made in the current era by mating puffins and ostriches, distant relatives of dinosaurs.[13]

We reached the river. It was running feverish from recent snowmelt. Tracks extended along both banks, so everybody rolled up their pant legs, removed their shoes and socks, and crossed the river. In one slippery spot between smooth boulders, the water was knee deep. This close to the mountains, the bottom was a mix of sand and cobble. Someone had checked a weather app, and the next snow, it now seemed, would hold off until we left the canyon. Getting soaked could mean hypothermia, nonetheless, so I took the crossing slow, and gave a shout after stepping onto the log at the far bank. When it was her turn, David's dog Cassidy balked, so he carried her across like a baby.

Paleontologists working with the US Forest Service had manicured the track site to expose more sedimentary layers. Today the pathways extend to almost the length of a football field. There are more than 2,100 individual footprints and at least 130 separate pathways in the rock. Eight hundred new tracks have been discovered since the 1980s. The brontosaurus tracks were elephantine and sunk so deep it was difficult to imagine how the giant herbivores didn't mire beside the shallow lake. Smaller bird-like tracks were made by the meat-eating allosaurus. A few tracks of the duck-billed camptosaurus have also been found.

I've plucked knuckles of Pleistocene horses and chips of mammoth tooth from prairie river sandbars. Holding evidence of their once corporeal realm, I could imagine those horses and mammoths running the steppe, because part of it survives in a form similar to the landscapes those extinct Pleistocene mammals wandered in the postglacial era. Russian paleontologists have unearthed mammoth meat so well preserved in permafrost that some have tasted it raw (though because ancient fats thaw out into a substance called "grave wax," nobody would want to eat mammoth for any reason besides culinary machismo). The dinosaurs of the Purgatoire inhabited a world far different than modern Colorado; but following their footprints made them seem more real than any dusty museum skeleton could. Lockley's work opened a new window into this long-vanished world. Like ourselves, what loped once lived. The six of us split up on both sides of the river to walk alone with the dinosaurs.

In the 1980s, Lockley led a *National Geographic* magazine film crew into the canyon. On the last night of their work, the crew mounted a camera on top of a fifteen-foot ladder secured against the wind by guywires, to get a drone's-eye perspective on the trackway. Lockley crouched next to the tracks and held as still as possible. During the four-minute exposure, one of the photographers— dressed in black to blend in with the darkness and remain out of the final print—walked around "painting" Lockley and the hundreds of dinosaur tracks with a flashlight. The technique reflected light from the flashlight, the river, and water in the depressions to form a stunning panoramic view. It looked like the dinosaurs had just finished their stroll beside the lake and might be somewhere around the bend nearby.

In the last few months before our visit, somebody had poured plaster into one of the allosaur tracks. Rangers had attempted to clean it out, but the chalky rubber still coated the depression. Damaging the tracks or any historical heritage site in Comanche National Grassland is illegal under the American Antiquities Act and other laws. The sites thus far have withstood the impact of visitors. The canyon is protected by its remoteness and lack of roads and amenities. To see the tracks you need to walk, ride a horse or mountain bike, or take one of the four-wheel-drive auto tours the Forest Service offers during summer. Crowds haven't stolen the ancient feeling of the place yet, perhaps because Picketwire Canyon is a grassland at the edge of the canyonlands, rather than a true desert canyonland like those enshrined as national parks and national monuments in New Mexico, Arizona, Utah, and southwestern Colorado.

As we began hiking back toward camp late in the afternoon, the low sun cast shadows from the canyon rim and concentrated the weak winter light. A sign in the prairie said: *TRAIL WILL BE OBLITERATED*. It stood beside an almost invisible part in the buffalo grass that we had missed on the hike out. Trail will be . . . obliterated. What about the trail required obliteration or hyperbolic signage? Contaminants, garbage dumps, unexploded ordinance?

The sign was a reminder that we hadn't found all the footprints of the canyon; there was another kind that had recently threatened to squash *all* the footprints of the Purgatoire River valley.

At some point in their journeys from source to mouth, the grassland rivers of America flow through or close to reserved federal lands of every jurisdictional persuasion, including national parks, monuments, forests, grasslands, recreation areas, battlefields, wildlife refuges, federally recognized Indian reservations, BLM lands, and facilities of US military branches. The military controls significant swaths of real estate—more than twenty-five million acres in the United States alone—including Fort Carson in southeastern Colorado. Fort Carson was established south of Colorado Springs on land donated by the state following the attack on Pearl Harbor in 1941. By the mid-1960s, the fort housed nine army divisions on 137,000 acres.

The army added 283,000 acres adjacent to the Purgatoire River in 1983 to create Piñon Canyon Maneuver Site, or PCMS, a large-scale combined arms training facility. This was the year when the made-for-TV movie *The Day After*, filmed mostly in Kansas, gave a generation of Cold War–era viewers apocalyptic nightmares about nuclear war with the Soviet Union. Although some ranchers sold their lands willingly, the army acquired half of the 350 square miles for the site by eminent domain. After efforts by Colorado Senator Tim Wirth and others, the army transferred Picketwire Canyon to Comanche National Grassland in 1991, to preserve the rare ecosystem and fragile dinosaur track site. The army made two promises when they opened the PCMS: there would be no live-fire maneuvers and no future expansions to the site.

In 2006, after troop numbers swelled at Fort Carson in response to base closures and new post–Cold War enemies in Afghanistan and Iraq, the army announced that it was studying a proposal to expand the PCMS by another 418,000 acres. The current maneuver site, they said, wasn't big enough to support the kinds of operations required to prepare troops for the new wars abroad.

That's what was in the press release. Leaked documents told another story. The new expansion was only phase one. The army eventually would require almost *seven million* additional acres, virtually all southeastern Colorado. The land grab would take the entire Comanche National Grassland and displace 17,000 residents. Local ranchers, already suspicious because the army had broken its promise when it started allowing up to .50-caliber machine gun fire at the site, were incensed at the new proposal. Some of them began to lobby federal and state officials, and anybody else who would listen, to look deeper into the effect the proposal would have on the valley of the Purgatoire.

The establishment of the PCMS in 1983 had been contentious at first. Former Las Animas County commissioner Mack Louden, whose family has ranched in southeastern Colorado for generations, remembers the antagonism between the army and ranchers. He told me, "I'd just returned from ten years in Montana where I had worked on some ranches. At the time, real estate developers working with the military were going ranch to ranch. They bought a fair amount of land, but there were some other people who didn't want to sell. They had to condemn the rest, and though there were court cases about how much the land was worth, in the end the army took everything by eminent domain."[14]

Over the following decades, the communities of southeast Colorado adapted to their new neighbor. The military conducted maneuvers for four months during the year. Besides the fact that the land west of Comanche National Grassland and the Purgatoire River was now off-limits, life went on as it had before. Many felt, and were proud, that families and local communities had done their duty for the country. There was a stir when, after twenty years honoring their pledge, the army began to conduct live-fire exercises.

Louden said, "Some people called them on it, but a spokesperson said: 'Show us the documentation.' Folks felt betrayed. That was back in a time when if you gave your word that was it. We didn't have to have everything written down. It was a different era." By the early twenty-first century, however, the army had become just another resident of the sparsely populated region.

Until 2006. In a move that rattled nerves from Trinidad to Las Animas, the army announced plans for the new expansion. Part of the reason was that Fort Carson, like few other major bases, had close access to the maneuver site. Louden said, "It was a couple hundred miles round trip from Fort Carson to the PCMS, but from the point of view of a general or a colonel, they could get in a helicopter, fly to the Piñon Canyon Maneuver Site, watch what's going on, and get back to Colorado Springs in time to have a drink that night. Fort Ord California is not quite the same."

The army wanted to triple the size of the base. It said the new lands would come only from willing sellers this time. Nobody would be forced to sell ranches that had been in their families for generations. But according to army planning documents, the military had a much larger long-term vision for the PCMS. The extra 418,000 acres would only be the first phase. To prepare for post–Cold War campaigns in the Middle East and Afghanistan, they wanted to expand the maneuver site to 6.9 million acres. They had already built a fake Iraqi village on the PCMS. Battalions could rehearse battles in an area the size of a small nation state. Louden said the eventual maneuver site would span "from Interstate 25, across the Arkansas River to the Kansas state line, and down to the New Mexico border," a box the size of Massachusetts cordoned off from the public and excised from the state of Colorado.

History was repeating itself along the Purgatoire. The deal echoed the 4-million-acre Las Animas Grant of 160 years before. Not only had the army reneged on its promise not to conduct live-fire exercises, now it was going back on its promise never to expand the PCMS. For many who remembered the 1980s, this went too far. Army planners believed the promilitary culture of the area would make it easier to justify another expansion. They underestimated how people would react when faced with the potential end of centuries of culture in southeastern Colorado.

In 1960, Robert Herrell opened a dental practice in La Junta and moved his family from the small town of Burlington, Colorado, at the time known for its hand-crafted Philadelphia Toboggan Company Carousel. (The carousel was damaged in 1981 by thieves who hacked off four of the menagerie's hand-painted rides during a severe thunderstorm—three horses and a donkey were later recovered in Kansas and repatriated to Burlington.)

Neither Burlington nor La Junta was a hotbed of 1960s student activism—hippies in conservative southeastern Colorado were a rare species. Herrell's teenage son Jim Herrell and a few of his high school friends, however, were fascinated by politics. They campaigned for Robert Kennedy's 1968 presidential campaign, got thrown out of an Edmund Muskie rally by federal marshals in Pueblo, and questioned the relevance of the Vietnam War, not because they were against soldiers or the military but, as the younger Herrell told me, "We didn't want ourselves or our friends dying in some place we couldn't pronounce."[15]

Jim Herrell became vice president of instruction at La Junta's Otero Junior College. Trained in special education, his other passion was the paleontology of the Purgatoire River valley. A well-respected amateur, Herrell worked alongside Martin Lockley and other heavyweights in the paleontology world, pushing for more recognition of sites in southeastern Colorado. "There may be more famous digs in Montana and South Dakota. But here in the canyons we've got a rich cross-section of fossils from the Jurassic, the Triassic, and, if you combine a few different locations, strata from most of the Mesozoic. It's that diversity of periods that sets us apart."

Although army researchers conducted archaeological and paleontology work on the PCMS, after it opened Herrell never ventured onto army land and worried about the scientific and environmental implications when the new expansion was announced. "Part of our thing with the army was natural resources," he said. "Some university scientists had discovered a site on a ranch south of here that gets clear back into the Triassic. It has early dinosaur-like animals, early mammals. We got the feeling that whole area was rich with undiscovered natural resources. The army wanted to take a staggering amount of southeastern Colorado for that base. There were natural resources of great importance being discovered, and obviously they won't be making any more of them."

At public meetings soon after announcing the expansion, the army argued that the larger site could benefit local economies. Its own reports, however, downplayed the possibility. Louden said, "Some local outfits tried to get contracts for construction, earth moving, porta-potties, that kind of thing. But most of the work was bid out to companies in Colorado Springs and Denver that knew how to get government contracts. They had given a few token things to the communities, but when the soldiers were down here, they stayed on site, they didn't go into the towns or to the restaurants."

Louden also distrusted the army's stewardship of the fragile high plains ecosystem. "A lot of that land had been plowed up during the Dust Bowl, it hadn't regenerated to original conditions, but it was in decent shape," he said. "When the army told everyone that now what's there on the PCMS is all pristine is a bunch of garbage. They were going in and reseeding land after they tore it up, and not using native species. Native stuff takes a long time to establish. You can plant a cover crop that takes hold quicker but it's not the right species for the land there. The entire Comanche National Grassland was going to be eliminated, even when they rolled the proposal back later. There were a lot of people who love the public lands, who regardless of how they felt about private ranches, didn't want to see public land converted into an army maneuver site."

When the army prepared to start purchasing land in 2007, some of the area ranchers who opposed the deal joined together to form a grassroots organization they called the Piñon Canyon Expansion Opposition Coalition. Herrell, Louden, and Jean Aguerre created their own group, "Not 1 More Acre," and sued the army in federal court, arguing that the environmental impact statement for the expansion was inadequate and that appropriations to acquire the land were illegal. US Senior District Judge Richard Matsch, who had presided over the Oklahoma City bombing trials, heard the case.

The army's environmental impact statement covered two main requests. One was a proposed expansion of the existing cantonment area to include new buildings: a brigade support complex, medical clinic, storage facilities, and upgraded roads and utilities. The other was an expansion of training operations on the expanded PCMS: it would require adding a live hand grenade range, an ammunition holding area, and upgrades to live-fire ranges.

Not 1 More Acre's lawsuit challenged the second request.[16] It argued that the environmental impact statement did not clearly specify how the changes would impact training operations, nor could they. Louden had seen firsthand what tanks do to prairie. "South of La Junta, you can still easily see the ruts of the Santa Fe Trail," he said, "and the trail hasn't been used since the 1800s. Imagine what tank tracks do. It's not like we're in Washington state with sixty inches of rain per year. The land can recover there in no time. But here, if we destroy what we have, it's not going to come back overnight."

According to the environmental impact statement from the original establishment of the PCMS in the 1980s, the army seemed to agree. The report said, "land in semi-arid southeastern Colorado cannot accommodate perpetual use for maneuver training."[17] The army was asking to expand the site so that damaged ranges would have more time to recover. Herrell, Louden, and Aguerre, however, believed the long-term economic, cultural, and environmental impact of converting all southeast Colorado, or even 493,000 acres, to a military maneuver site eclipsed this short-sighted argument. Louden believed it would end a long history in this spare cultural crossroads. He said, "We value our culture here and if they take over southeast Colorado, that culture is gone."

The lawsuit went to court in 2008. After closing arguments, Judge Matsch returned a verdict that stunned even Herrell, who always believed they had a strong case. In his decision, Matsch wrote that the proposed expansion of PCMS would permit the entire site to be used for training purposes every day of the year and that the army's conclusion that there would be no significant environmental impact was counterintuitive. It was obvious that such intensive use of the PCMS would prevent any meaningful mitigation of the resulting environmental impacts. The army had failed to develop a use model that would allow for the rest, recovery, and restoration of fragile lands near the Purgatoire. He refuted the army's plan to change how it conducted maneuvers on the PCMS and ruled in favor of Not 1 More Acre. Five years later, in November 2013, US Senator Mark Udall of Colorado and Assistant Secretary of the Army Katherine Hammack held a press conference in Pueblo to announce that the army had dropped plans to expand the site and would withdraw its request for a land acquisition waiver.

Today, soldiers from Fort Carson still conduct trainings on the PCMS, preparing American forces for battles in arid lands abroad. Paleontologists continue to make discoveries in their ancient digs. The Santa Fe Trail is still visible on the high plateaus south of La Junta in Comanche National Grassland. Herrell told me he was proud of the work he did on behalf of this remote and hidden landscape. "Don't ever think for a minute that a small group of people can't make a significant difference. Today everybody is so cynical and siloed. But we had ranchers teaming up with the Sierra Club. I mean conservative NRA ranchers. In America, a small group of people can still make a significant difference against what looks on paper like insurmountable odds. The three of us took the secretary of the army and the secretary of defense to court. And we won."

Although the army has no current plans to expand Fort Carson or the PCMS, Herrell and Louden both said they're not sure how long that will last. Given political exigencies and the increasing price of land in Colorado Springs, the army could one day reconsider. Herrell said, "Someday in some little earmark of some little bill, a senator could sneak in an authorization to expand the PCMS again." Louden said, "I'm guessing sometime in the next ten years, after enough people forgot about the last time, they're going to try again."

Jim Herrell told me he hoped that, if that day comes, the people of southeastern Colorado will again stand up for what they believe in. "The point is that our constitution and our government only work when we're all active participants."

On the hike back, we explored a boulder field and found a linear pictograph called snake glyph. It looked more like a map out of the canyon than a snake, perhaps a cartographic warning from the past that this part of the Purgatoire valley is a place to visit, not remain.

The next morning in La Junta, we ate breakfast in a café owned by Larry Tanner, a fourth-generation Coloradan whose father lost a 1952 bid for congress running on the motto "Don't make our future any darker than it already is." All nine of his siblings—whose school pictures are prominent parts of his father's campaign poster displayed behind the register in the restaurant—still live in the area. A waitress with a nose ring greeted customers in Spanish and English, and a man dressed in a POW/MIA jacket, carrying a full-sized American flag on a pole, came in and ordered scrambled eggs with green chile.

Hikers, mountain bikers, bow hunters, mountain biking bow hunters, paleontologists, tarantula lovers, and others share Picketwire Canyon with the antelope, mule deer, prairie dogs, porcupines, road runners, and cattle that graze federal permits. As ancient footprints attest, the Comanche National Grassland and Piñon Canyon Maneuver Site are only the latest in a long series of newcomers to the valley of this rare grassland river of the canyons.

A light snow began to fall on the drive out from La Junta. Near the ghost town of Arlington, a north wind started nudging my car toward the shoulder. A golden eagle manned its solemn post on a utility pole while its mate soared above a field of whitening cornstalks.

There is no end to observing nature.

Already, I was thinking about the South Platte, another river of thirsty plains. That trip would have to wait for my knee to heal, and for the spring melt to raise river levels high enough to wet the gunwales of my canoe. Although it shares characteristics with its siblings to the south, the South Platte is a river of the Front Range, embodying all the challenges and chagrins of the urban-rural divide and the waning waters of the West. They may share a state, but the South Platte flows through a much different Colorado than does the River of Lost Souls.

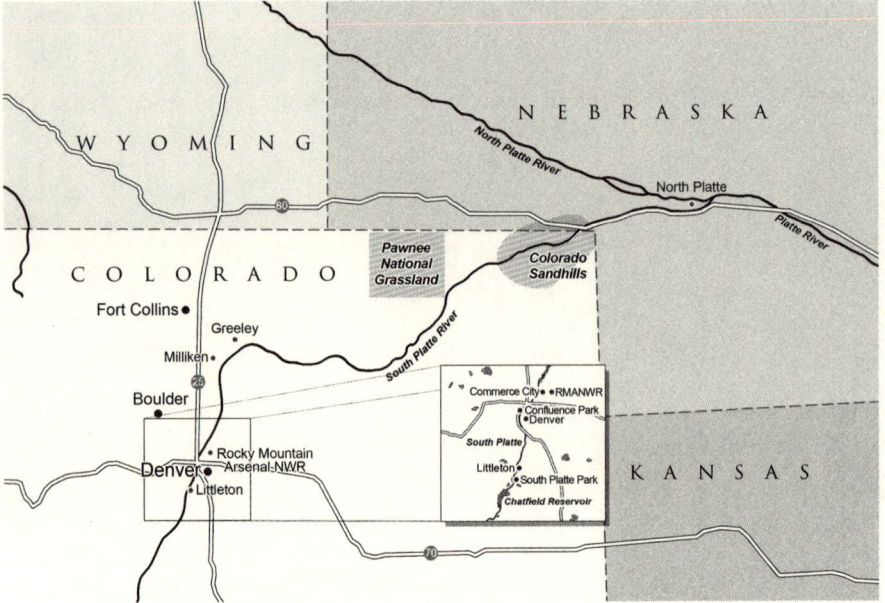

MAP 6 The South Platte River

CHAPTER 6

SOUTH PLATTE

We came to the shallow, yellow, muddy South Platte, with its low banks and its scattering flat sand-bars and pigmy islands—a melancholy stream straggling through the centre of the enormous flat plain, and only saved from being impossible to find with the naked eye by its sentinel rank of scattering trees standing on either bank. The Platte was "up," they said—which made me wish I could see it when it was down, if it could look any sicker and sorrier.

MARK TWAIN, *Roughing It*

Environmental crimes, civic neglect, even a signature style of literary disrespect. The South Platte has suffered indignities since long before Mark Twain inked his ill regard for this high plains river one hundred and fifty years ago. It is the principal tributary of the Platte proper, the storied Nebraska crane river where a half-million sandhill cranes congregate each spring in one of America's last great migrations. But the South Platte is a river with its own history apart from any downstream destiny. Even with a grassland twist, it suffers all the contemporaneous problems of western rivers, as it ferries snowmelt from the Continental Divide out into the parched shortgrass prairies of eastern Colorado and southwestern Nebraska.

The remaking of Denver as capital of a dominant new western persona—The Front Range!—has not left the South Platte behind. Perhaps no city of the grasslands is more dedicated to the vitality of its landmark river than Denver, a city that is also home to the largest urban prairie rewilding project in America. But explosive growth along the Front Range and an insatiable thirst for scarce water resources have created challenges, challenges that could worsen as warming temperatures assault timeworn hydrologic patterns on the eastern slope of the Rockies.

The story of the South Platte begins as a story of meltwater. Originating in glacial basins and tarns below the high peaks of the Mosquito Range near Breckenridge, the river flows four hundred miles across the central Rockies, through the heart of metropolitan Denver, along the Colorado Piedmont east of the Front Range, and northeast across shortgrass prairies and the Colorado Sandhills before reaching Nebraska. Every year in late March or early April, sunshine and warming temperatures begin to soften the snowpack in the central Rockies. This lubricating metamorphosis transforms snowfields into rivulets along the backbone of the Continental Divide. The rivulets eventually coalesce into the main tributaries of the South Platte: the Cache la Poudre and Big Thompson Rivers, St. Vrain, Boulder, Clear, Tarryal, Sand, and Wildcat Creeks, and the South and North Forks of the South Platte.

Before climate change began pinching mountain snowfall (according to the US Environmental Protection Agency, between 1955 and 2022, April snowpack levels dropped across the intermountain West an average of 23 percent[1]), the twenty feet or more of snow that fell each winter started to melt by early April, gorging streams at lower elevations. As the melt intensified over the next eight weeks, the synchronized chaos of fluvial geometries produced peak flows during the second week of June. A seasonal tide of mountain water poured down from the peaks into the prairie, spreading in a plume across northeastern Colorado until the river joined forces with its meltwater sibling the North Platte in western Nebraska. There, the two rivers form the main stem of the Platte, the 310-mile river that meets the Missouri near Omaha. After the drama of the melt crescendoed, the South Platte sank back into its banks, and in some years disappeared altogether into the hot sands of the prairie. Not all the snowmelt reached Nebraska. In high-elevation grasslands east of the mountains, some of the melt evaporated, some disappeared underground in subsurface channels, and some seeped deep into sponge-like rock aquifers buried beneath hundreds of feet of soils that formed when the mountains wore down over millions of years. Like its patron state, Colorado, the river has a history of boom and bust.

Hydrological records from the late nineteenth and early twentieth centuries show the timing of the melt was consistent year after year, but today, a warming climate is affecting the depth of the snowpack and the timing and rate of the runoff. This is having a profound effect on the river and the native grasslands that have managed to survive in its watershed. Even the effects of global climate change, however, pale when compared to the more immediate demands of an exploding population along the Colorado Front Range. Flood control and the complex parceling of water for residential consumption, industry, and the irrigated cornucopia of the "fruited plain" all affect the river as it undulates across a landscape that in recent geological history was Sahara Desert dry. The tale is a familiar one—a grassland instance of the water resource woes now playing out along the rivers of the water-starved American West.

By statistical measures of demography, suburban sprawl, employment, and sheer traffic gridlock, Denver may be the most transformed western metropolis of the twenty-first century. The South Platte, its keystone river, is both beneficiary and victim of this makeover. The city was once known for memories of the Old West and their twentieth-century echoes: the Pikes Peak gold rush, Buffalo Bill's grave, a government mint that produced pennies stamped with a "D," Coors beer, sooty air from industries that poured smog into a self-contained bowl of mountains, ski bums down for a breather, vacationing beatnik-wannabees chasing the ghost of Neal Cassady along Colfax Avenue, Mile High Stadium, and John Denver. But a booming twenty-first century economy, complete with high-paying tech jobs and the promise of fun in the outdoors, has created a "new Denver" that is more Front Range than Old West: legalized marijuana and pot dispensaries (Coloradans passed a state constitutional amendment that legalized recreational marijuana in

2014), hipster neighborhoods, a thriving foody culture, art districts, hundreds of miles of bike trails, and an almost cult-like devotion to craft beer (Colorado had the fourth-most breweries per capita in the United States in 2024[2]). Compared to most western cities, Denver is looking toward its future rather than its past for its identity.

The mountains have always lured people to Colorado. But Denver, the state capital, is on the Great Plains, *not* in the mountains. The city owes its existence, both past and future, to the South Platte, a river undergoing its own urban revival. Indeed, Denver and the communities in its close orbit might be the strongest municipal advocates for grassland river revival in the trans-Mississippi West.

Driving south from the capital city toward Littleton at 6:00 p.m. on Friday night, we felt like herd animals packed into the hydraulic chute of Interstate 25. Next to the highway, two trains passed each other traveling in opposite directions. One pulled a line of coal cars, the other a line of low trailers, each carrying the gargantuan blade of a wind turbine, an almost-too-good-to-be-true metaphor for modern Denver. We were searching for the precise spot near the Mineral Avenue bridge where the South Platte transitions from a mountain to a grassland river. Parking at a ride share station near the Carson Nature Center, we walked the streamside trail and scrambled down an embankment to the river's edge. Beside a sandbar covered with cottonwood sprouts, the thin current flowed across a span of hand-sized boulders notched in the middle with a chute. A mother mallard urged her brood of four ducklings into the shade of the bridge. Water bubbled over a small rapid and slowed as it washed over the first real stretch of sand, one of the hallmarks of prairie streams and evidence that the river was transitioning from cobble to a sandy bottom.

The stream gauge below Chatfield Reservoir was holding steady at 240 CFS, rain in the forecast.[3] We drove farther upstream and my friend Chris helped me launch my canoe above two "guide boulders" that marked a small rapids, if you could call it that. Despite

FIGURE 6.1 South Platte River during peak melt in Littleton, CO. Photo by the author.

the modest flows, the rapids doused me and I was soaked for the remainder of the short paddle down to Reynold's Landing, a popular spot for tubers who want to get on a river without leaving the Denver metro. People who had never paddled it themselves had tried to warn me off, for fear of foul smells and *E. coli* bacteria, but as I dragged the canoe around the chute at Mineral Avenue and slid back into the river, a yellow warbler flashed across the water and landed on a staked cottonwood sapling alongside three western kingbirds. The ensemble seemed ready to pose for a bird-book photo shoot. The ninety-minute run was pretty and peaceful, and it ended for me at sunset beside a riverside restaurant full of Denverites settling into the weekend. Until recently, building a restaurant or any establishment next to this river would have been unheard of.

Until the mid-1960s, the South Platte was undammed in the Great Plains, although in the previous century a series of reservoirs had been built along its largest tributary, the South Fork. Fresh from the mountains, after the river reached the Denver metropolitan area it started showing up on lists of the most debauched urban streams in

America. In addition to chemicals such as sulfur, arsenic, uranium, lead, cadmium, and zinc from placer mines in the mountains, cities along the Front Range and adjacent farms contributed high concentrations of nitrogen, potassium, phosphorus, insecticides, pharmaceuticals, raw sewage, fecal coliform, and *E. coli*. Denver walled off the South Platte near downtown with industrial facilities and dumps. Twenty years before Mark Twain saw the river, the historian Francis Parkman wrote that "the South Platte is a sorry river . . . so dirty that it makes one's flesh crawl to look at it, and it smells like a slaughterhouse."[4] Some twentieth-century Denver residents didn't know the city *had* a river.

Intense thunderstorms containing hail, tornadoes, and gustanadoes—the sensuous horizontal wind tubes that roll down mountainsides before righting themselves and attaching to low cumulus clouds that fuel further rotation—are common in eastern Colorado. In June 1965, the Front Range was pummeled by severe weather. On the afternoon of June 16, storms parked over the plains, sending the river frothing over its banks and into the neighborhoods of metropolitan Denver. A year's worth of rain fell in a single workday.

I asked Skot Latona, who is a heritage interpreter, prairie plant expert, and manager of South Platte Park, the 880-acre linear nature preserve in Littleton that surrounds the wildest remaining stretch of the river in the Denver metro, about the flood. He said that people stood on the banks and watched as a twenty-foot-high, mile-wide wall of water and mud, gargling mobile homes, semis, propane tanks, and garbage from landfills, scoured the bottoms in Littleton and Centennial. "The drainage area of the river all the way to South Park and Breckenridge is massive for a river this size, and on top of that, there is the volatility of Front Range thunderstorms. We get rain events that stall out and dump ridiculous amounts of water. In 1965, Littleton had its turn. The water formed into an ocean wave a mile wide coming down the valley."[5]

When the wave reached Denver, the river regurgitated the pollution that had been dumped into the channel for decades, inundating the city in one big filthy surge, taking out bridges, swamping power stations and chemical plants, killing twenty-one people, and inflicting half a billion dollars in damage. It was one of the worst natural disasters in Colorado history. The flood catalyzed the rapid construction of Chatfield Reservoir, part of a floodwater control project approved years earlier. South of the city, Chatfield was just the latest man-made modification of the river. Littleton had been founded near an irrigation inlet canal off the South Platte. The canal was later repurposed into a flour mill that operated for eighty years. During the nineteenth century, thirty more irrigation ditches were dug to siphon water into crop fields. Ten are still in active use today.

Although the reservoir became a popular spot for sailing, water skiing, swimming, and fishing, city officials in Littleton pushed back when the US Army Corps of Engineers released plans to turn the river into a big drainage ditch below the new lake. Latona told me, "After they finished Chatfield, the Corps of Engineers analyzed the river downstream through Denver. They planned to carve a straight trapezoidal channel for the river to run through lined with rock and concrete—no connection to the floodplain, no woody vegetation, no meandering whatsoever." City officials in Littleton didn't want to divert the river, which ran through the heart of their town, into an open-faced concrete pipe. They lobbied the US Congress to get permission via the 1974 Water Resources Act to purchase land in the floodplain. It was "650 acres at first," Latona said, "and eventually it grew to 880 acres. The intent was to surround the river with a natural floodplain to absorb waters if Chatfield ever releases due to some extensive rain event."

Littleton created not only a park and nature preserve but a "river within a river" that mimicked the natural sinuosity of the South Platte prior to damming and channelization. "We rebuilt the channel for the two and a half miles it runs through the park. It includes a low flow channel with an approximation of the original braids the river followed with islands and sandbars that are in place whenever the river runs less than 650 cubic feet per second. We can't re-create the historic floodplain or raise the channel back up. But we've created some concentrated flows within the channelized river that help with water quality, water temperature, and that also have an aesthetic appeal."

South Platte Park is one of several river-themed parks in the Denver metro. One of Latona's missions is to restore native habitat where possible, given urban interruptions like roads, bridges, and parking lots. He said, "We've reestablished a lot of native grassland, prairie plants, native wildflowers, and cottonwoods on the main course." The streamside vegetation and functional channel have improved water quality. He added, "Until about the late 1980s, the South Platte was literally considered the sewer of Denver. There was untreated waste and untreated chemicals. The riprap that many private landowners used was old automobiles. It was a junkyard with very poor water quality. We have one wastewater treatment plant upstream of Littleton below Chatfield, and obviously they are releasing water that meets all federal and local requirements."

About the same time Littleton began to revitalize the river below Chatfield Reservoir, officials and private groups in Denver were looking for ways to reclaim the industrial wasteland surrounding the river that had caused big problems in the flood of 1965. When state senator Joe Shoemaker ran for mayor, he campaigned on a platform that included river restoration. Although he lost the election, the person who defeated him, Bill McNichols, never forgot Shoemaker's passion for the South Platte. In 1974, McNichols asked Shoemaker to draft a plan to "take back the river" and backed it with two million dollars in funding. Shoemaker eventually chaired the Platte River Development Committee, the principal advocacy group for the South Platte River in Denver. Today it is called the Greenway Foundation, led by executive director Ryan Aides. Aides told me, "Shoemaker didn't have a specific mandate. At that time there were no parks along the river. Not a single greenway or trail. There was nowhere you could interact with the river, take your dog, have a picnic."[6]

The lack of access was the first problem Shoemaker dug into. As other river advocacy groups had discovered, few people care about rivers if they can't experience them or don't know they exist. Aides said, "Joe's vision was that if you brought people to the river, they would invest in it, on a personal level. He said, 'One day, the best place to live, work, and play in Denver will be along the banks of the South Platte River.' People laughed at him. They called it 'Shoe-maker's folly.' Everybody thought he was crazy."

Shoemaker's group targeted some of the nastiest, dirtiest dumps along the river—landfills, dump spots for crushed cars that once qualified as "bank stabilization," paint factories—and turned them into parks. Globalville Landing Park and Johnson Habitat Park, once landfill dump sites, were first, but the jewel of the revitaliza-tion became Confluence Park, an old energy substation beside a capped landfill. It was a strategic spot to celebrate the river and Denver. In 1858, gold was discovered at the confluence of Cherry Creek and the South Platte, leading to the founding of the city. The original land use plan for Denver included land near the confluence for a major urban park, but it was never built, and soon the conflu-ence got swallowed into the urban core of the city as warehouses and industrial plants were built up around it. Aides said, "From the city's perspective, this is where everything started, this was the birthplace of the city. So, this was where we started reconnecting people to the river."

Today, the Confluence Park area is called Shoemaker Plaza (Joe Shoemaker died in 2012). It includes a kayak run, sand beaches, bike trails, a concrete promenade, and is adjacent to a redeveloped warehouse district with pedestrian bridges, loft developments, and a large outdoors store. All this activity brings hundreds of people to the river on an average weekend night, many fishing waist-deep or surfing the engineered riffles above the park. Boulder gardens and whitewater spots throughout the city have created an ardent community of stand-up paddleboarders and river "surfers" who hold gear swaps with skateboard and snowboarder groups, do river cleanups, and sponsor events at popular standing waves. Conflu-ence Park may be the single-most utilized spot on any grassland river in the country.

The greenway bike and running trails are the additions that have brought the most people down to the river since Shoemaker's revi-talization efforts began. The trails connect all the river towns for miles. Skot Latona told me, "You can take the trails from Chatfield Reservoir all the way up to Adams County and never cross a road.

You're on the river the entire way. I've been in cities where people can't figure out how to get to the river. Today, we have thousands who visit some stretch of it every week as part of their daily routine. Hikers, runners, and bicyclists can follow the river almost thirty miles from Brighton south to Englewood, through Denver's urban core, industrial districts, the River North Art District, lower downtown, then on the Mary Carter Greenway south toward Littleton and Chatfield."

As the river has become cleaner, paddlers have also returned. Latona said that twenty years ago in Littleton, "the only people on the river were kids who tubed during summer camp. We'd get twenty kids at a time at the peak, and typically it would be more like three or four per day. Since then, it's taken off. We see big groups coming through the park and people using their own boats. We hire a parking management company at our put-in and last year we had seven hundred people float through our section in one day."

As with every usage of the South Platte, however, the challenge for paddlers is getting enough water. When Chatfield Reservoir is not releasing—common in summer—the river runs below the 100 CFS minimum flow needed so paddlers can avoid damaging the fragile restored bed of the river. Latona said, "In recent years, it's become a challenge to get the minimum flows. Nine of ten years, the river is completely controlled by releases from Chatfield. Last summer, there was so much rain in the upper basin that Clear Creek, up in the mountains, closed tubing because the river was so high. But we didn't see any of that water down in the city, on the plains. Runoff only affects us in a super wet year, or when we get meltwater from more than a 130 percent snowpack."

Ryan Aides agreed and told me that, beyond water for recreation, river levels also affect water quality. "In Denver, one of the biggest obstacles is the lack of water in the river during the summer. The river can run so low that trash builds up. It increases *E. coli* and nitrogen levels. They say pollution is a problem in search of a solution by dilution. If we could get more water, it would help with water quality, but this is not just an issue for Denver, it's a problem across the entire American West."

Denver and Littleton can't just place an order for more water from one of the reservoirs that impound the South Platte: Antero, Spinney Mountain, Chatfield, Eleven Mile, Cheesman, or Waterton Canyon / Strontia Springs. In all but the rainiest years, the water that flows through the cities is a carefully allotted and managed resource, parceled using "intentional transfers" delivered to downstream users who place orders based on a complicated history of water allocation practices and laws. Allocations can change week to week, day to day, or multiple times in a single day. At the heart of water allocation on the South Platte, the Colorado, and rivers across the American West, is a queue—like the beer line at a Rockies game when they're winning—only this queue began a hundred and forty years ago and has only grown longer since. Understanding exactly how this queue works is crucial to understanding water policy in the West.

The Colorado Doctrine, or "doctrine of prior appropriation," is a biblical-sounding name for a collection of laws and agreements that govern water allocation. Colorado prairies are dry—almost, though not quite, as dry as true deserts. From the time the Colorado Territory was established in 1861, water became a central point of contention, legislation, and litigation. From the beginning, elected officials and courts presumed that the purpose of water was to use it. Water laws are complex, but the problems they aim to solve are simple to state. The first laws were adopted to govern water access for mining. No matter how they might swerve and dither, the geometry of streams is linear. Miners used water to separate slurry from silver and gold. A claim without water had little value.

Suppose you stake a claim next to a creek. In theory, another miner could start working higher up on the same creek, divert all the water, and leave you dry and penniless. To avoid such upstream diversions, miners created an at first unwritten creed: "first in time, first in right." The earliest claims to water, temporally speaking, would be the first to receive water each year. The oldest claimants couldn't take all the water, but they could get in line first for their portion. The arrangement eventually was applied to other types of water claims as well, especially for irrigation. This became critical as entrepreneurs during the Manifest Destiny era attempted to make the "Great American Desert" *bloom*. The only way that could happen in the high plains was with irrigation. Beginning with the Homestead Act of 1862, new arrivals intent on row crop cultivation in a dry land began digging irrigation canals to divert the South Platte and other Great Plains rivers.

Even today, there is no single set of federal water rights laws. Colorado began working out its own guidelines in the late 1800s. The Colorado Doctrine consists of four rules that direct every drop in the South Platte River: (1) The state's surface and groundwater is a public resource intended for "beneficial" use by public agencies, private persons, and entities; (2) A water right is a right to use a portion of the public's water supply; (3) Water rights owners may build facilities on the lands of others to divert, extract, or move water from a stream or aquifer to its place of use; and (4) Water rights owners may use streams and aquifers for the transportation and storage of water.[7] Wyoming, New Mexico, and Montana later adopted the same model of water appropriation.

The doctrine applies to rivers, lakes, ponds, and ditches, and to cities, counties, businesses, and individuals. When water rights holders want to use water, they place an "order" with the river commissioner. Based on the orders of the day and where water is stored (it could be in any of the upstream reservoirs), the water for the order gets released. The oldest claim will get all its request until its allocation runs out, and then the next oldest claim will get all its request, and so forth until there's no more water. Paddlers and tubers can float on water that is "owned" by a claimant, but they can't control when or where that water will flow.

The "last mile" of delivery usually takes the form of an inlet canal, like the pipe that connects a city's main water line into an individual home. One example is the North Sterling Inlet, a canal that diverts water from the river into North Sterling Reservoir in northeastern Colorado for storage. According to data gathered by the

Open Water Foundation, the inlet canal that connects to the reservoir is sixty miles long and includes a concrete dam and four radial gates where it meets the river.[8] Like most storage facilities on the South Platte, North Sterling Reservoir is "off channel." Instead of damming the main stem of the river, the inlet canal delivers water from the river to the 80,000-acre-foot capacity reservoir. The water district begins to fill the reservoir in October. Water for the most senior of rights holders is stored first. Once all the water rights filled from Sterling Reservoir are deposited in the lake, the inlet from the river is closed and priority is transferred to the next lake in the system, Prewitt Reservoir.

When a water rights holder orders water, the allotment is delivered from a sixty-five-mile outlet canal connected to the reservoir. North Sterling serves approximately seventy-five individual rights holders within the water district. The outlet canal consists of a concrete headgate, a steel slide gate, and a Parshall flume, the metering device that measures irrigation flows to audit the delivery of orders. A "turnout" is an event where water is sent from the reservoir into the canal and can serve multiple orders. Water requests usually run during the growing season, between April and September. Once water reaches a farm, it is spread either through flood or sprinkler irrigation methods. Flood irrigation is an ancient technique that relies on gravity. Sprinkler irrigation is a newer method that requires sprinkler devices and energy from fossil fuels or electricity.

The challenge with water allocation on the South Platte, like other western rivers, has always been scarcity—in most years demand exceeds supply. With population growth in the Front Range and decreasing snowmelt from climate change, the problem is getting worse. Over-appropriation isn't a new phenomenon. The river was overbooked as early as 1886, ten years after statehood. In his 1905 book, *The Conquest of Arid America*, William E. Smythe wrote that "the South Platte was always a doubtful stream, and its waters have been diverted in so many directions that there is scarcely enough left to make a decent-sized river."[9] Different schemes have been employed to increase supply, including construction of the Grand River Ditch in the early twentieth century. The diversion project shunts water that falls on the western slope of the Rockies in the Never Summer Mountains—water that should flow toward the Pacific Ocean—into a ditch that drains to the South Platte basin and eventually the eastern plains. Since then, more tunnels and pipelines have been built to divert water, apportioned by gravity to the Colorado River, into the South Platte, further bolstering supply to the eastern slope. Shortfalls are projected to worsen and, to account for increasing demand in coming decades, new projects have been proposed. One multibillion-dollar proposal described in a report by the South Platte Regional Opportunities Water Group would recapture water lost after transfer from existing diversions from the Colorado River.[10] The plan includes collecting agricultural runoff from farms in the South Platte basin and pumping it back for use in Denver. The reason it would go back to Denver would be to provide water for municipal rather than agricultural uses. In recent decades, cities have been buying farms with senior water rights as a way to control their own water supplies.

Grasslands used to be a factor in the hydrology cycles of the South Platte, but they play a lesser role today since much of the shortgrass prairie has been plowed. One grassland rewilding project outside of Denver, however, has already had a beneficial effect during a significant flood. Twelve miles from Confluence Park and the heart of downtown Denver, the grassland is one of the most unlikely prairie rewilding projects in America. Like river revival in the Denver metro, the project shows that meaningful grassland restoration can occur even in urban settings.

❧

Chloe and I huddled together in a pool of morning light beside a small hill that glimmered with broken quartz and red ants, indolent from a week of frosts that hadn't yet driven them underground to hibernate or perish. Raising her binoculars to a white dot in the outstretched arm of an arthritic tree, she mouthed the word "*porcelain*," confirming in our secret language of bird identification that the black-and-white smudge in the distant limb was a bald eagle.

Perhaps irritated by her quiet satisfaction, prairie dogs scolded us like killjoys from the adjacent hills.

This was a favorite trek we'd taken many times. Before the sun rose, we'd spotted an antelope, a pair of ferruginous hawks, two rack-laden mule deer mulling a fight, the eagle, and a burrowing owl that disappeared into a prairie dog tunnel like whack-a-mole when it noticed us. Chloe took out her phone for a selfie, not against a backdrop of yellow plains dissolving into the far horizon, but against the unmistakable buildings and mountains that define the skyline of downtown Denver, a jarring reminder that one of the most successful urban rewilding projects in America—Rocky Mountain Arsenal National Wildlife Refuge—is only twelve miles from the heart of the fastest growing metropolis of the West. The South Platte River runs parallel to the refuge a few miles to its west. In the Colorado prairies, bald eagles were originally river birds that overwintered in colonies, but breeding pairs also nested in riparian groves during the summer. A more contemporary convocation of the birds was crucial to the origination of the refuge.

Most of the Denver metropolitan area was originally a vast short-grass prairie. The most obvious reminders of the city's grassland past still scamper across vacant lots and green spaces throughout the eastern part of the city. In addition to three million human residents, the Denver metro is home to the world's largest urban population of prairie dogs. Nowhere is that more evident than Rocky Mountain Arsenal National Wildlife Refuge.

Beginning with indigenous peoples thousands of years ago, Denver has long been a transportation nexus. People followed the rivers, and the South Platte was the principal thoroughfare across a swath of shortgrass plains north of the Arkansas Valley. When the United States entered World War II after Pearl Harbor, the military began looking for strategic locations to manufacture supplies for the war. They chose a site to build what became Rocky Mountain Arsenal on a high grassland above the river northeast of downtown. Located in the heart of the country and close to rail lines and highways, Denver was a natural place for such a factory because bombers landed there frequently in the days when the big planes couldn't cross the country without refueling. Construction of the arsenal began in 1942 and, within six months, the plant was producing mustard gas, bolstering agents, incendiary bombs like the ones used in the firebombing of Tokyo, and sarin nerve gas.

After the end of the Korean War, the army leased the site to the Shell Oil Company to make agricultural pesticides and herbicides, while the military continued producing rocket fuel and disposing toxic waste on arsenal grounds. When military and commercial production ended in 1982, the soil and groundwater were contaminated with some of the most toxic chemicals of modern industry. Funds had to be allocated for a massive cleanup that focused on a six-mile core area where weapons and agricultural chemicals had been produced. When the plants were operating, they had pumped waste and byproducts into fill basins, but the poisonous plumes were never fully contained. In 1966, attempts were made to inject the wastes deep into the ground after contaminants were detected in the South Platte. The project was scuttled when Denver experienced earthquakes blamed on the injection wells. The arsenal eventually qualified for money from the Environmental Protection Agency's Superfund. By 1996, a remediation plan was completed, removing contaminated topsoil down to a depth of ten feet. The contaminated soil was hauled away and reburied in two hazardous landfills at the arsenal. The earlier waste disposal pits were too poisonous to remove, so they were sealed off to prevent groundwater and burrowing animals from infiltrating the soil. In some areas, contaminated groundwater was pumped out and filtered. Once the work was complete, a long-term sampling plan was implemented to monitor the site.

Among other products, organic chlorine pesticides like DDT (described in Rachel Carson's pivotal book, *Silent Spring*[11]) had been produced at the plant. These insecticides found their way into the bloodstreams of birds. In the 1950s and 1960s, raptor populations declined rapidly in the United States. The chemicals caused the shells of eggs to thin and break in the nest. Bald eagles, America's emblematic bird, wound up on the endangered species list and nearly went extinct, reaching a nadir of 417 individual birds in the lower forty-eight states at one point.

In 1986, after the arsenal had been closed to the public, a biologist located a bald eagle roosting in a tree in one of the contaminated sectors. If European brown bears, wolves, lynxes, and wild horses could return to the Chernobyl exclusion zone thirty years after the world's worst radioactive disaster, then the hardy animals of the shortgrass prairie could repatriate a high plains chemical dump. The US Fish and Wildlife Service investigated and found a communal roost of eagles, along with thousands of prairie dogs, mule deer, and pheasants. The eagles had created an opening for the return of nature.

Because bald eagles were protected under the Endangered Species Act, the army, Environmental Protection Agency, Colorado Department of Health and Environment, and Shell Chemical agreed to convert the site into a wildlife refuge after completing the cleanup. Environmental groups like the National Wildlife Federation, Denver Audubon Society, and local communities got involved, providing input on the long-term use plan. The 15,000-acre Rocky Mountain Arsenal National Wildlife Refuge, one of the largest urban national wildlife refuges in the country, was designated in 1992 and opened to the public in 2004. Managed by the US Fish and Wildlife Service, the refuge is adjacent to Commerce City and offers sweeping views of downtown Denver and Denver International Airport. By 2024, the refuge was home to bald eagles, golden eagles, ferruginous hawks, burrowing owls, badgers, mule deer, coyotes, bison, 45,000 prairie dogs, and black-footed ferrets, the most endangered mammals of North America. It's also one of the most aggressive prairie rewilding projects in the high plains.

David Lucas, who manages Rocky Mountain Arsenal and the other US Fish and Wildlife refuges in the Front Range, told me that prairie restoration is a key focus of their work. "Planting prairies started during the cleanup. Because they used disposal basins in the fifties and sixties, contaminants seeped into the groundwater, and it also allowed for airborne spread. When the wind blows, it blows water off the top of ponds and lakes, and that spreads contaminants into the topsoil, where it wasn't there before."[12]

As the topsoil was razored off, the small fragments of native prairie at the arsenal were destroyed by the cleanup. Lucas said, "At first, they planted nonnative grasses throughout the entire area after the topsoil was scraped off. But later, soil scientists, range technicians, and range specialists came together to look at the soil, which dictates what kind of plant communities should be there. They mapped the entire site and created plots to tear out the nonnative and invasive species and plant new native species. Those sites were plowed again and planted with sterile sorghum." Sterile sorghum is an annual that can't reproduce and is often the first species planted to stabilize soils. Lucas said, "After the sorghum was knocked down, they drilled in the native seed mixes, again according to soil type, and irrigated it for several years."

After that, Lucas explained, the focus shifted to managing invasive species, as it did on so many other prairies. Lucas told me, "We take invasive species management very seriously. We'll be fighting invasive weeds for the next fifty years. We use a program called integrated pest management that tries to find the least impactful method first. But it's a complement of biological controls, herbicides, manual setting of fires, digging, cutting, and mowing. We employ the entire range of techniques at the Rocky Mountain Arsenal."

Like the American Prairie Reserve and Grand River Grasslands restorations, metrics tailored to individual parcels are being incorporated into refuge planning. "Our measurement strategy is still under development," Lucas told me. "We've started implementing a grassland monitoring program. One of the big changes was when we started looking at larger patch sizes. At first, we were doing things twenty acres at a time. But then we stepped back, and started mapping things out based on fire, on holding lines such as roads to create larger tracts, anywhere from one to three thousand acres," he said. "Now we're targeting these larger tracts with similar characteristics instead of going in piecemeal. Although it's evolving, we've found we have two basic types of prairies: traditional shortgrass and mixed-grass prairies. We have management criteria for each type, and measure accordingly. The focal species are slightly different in the two types."

Preserve biologists also focus on grassland birds. They study species such as grasshopper sparrows and lark buntings and try to build up populations by bolstering shrub canopy for nesting and open space for foraging. Higher ecosystem-level goals are farther off. Lucas said that, while "thriving contingents of grassland species are our ultimate grassland metrics, . . . at this moment in time, to be brutally honest, it's about percentage of invasives. Cheatgrass and semiannual thistles are some of the worst." Cheatgrass, or downy brome, and semiannual thistles are prolific breeders that bioblast pastures across the Great Plains.

Lucas said that reintroducing bison has provided the biggest payoff. Starting with twenty animals, the herd has grown to between sixty and eighty bison on a limited section of the refuge. With adjustments, Lucas believes they could expand the herd into the hundreds. "Bison are eating our native grasses at the right times. They are trampling, they are creating nitrogen cycles. They're doing things we cannot do. They eat forty pounds of grass per day," Lucas said. "They don't loiter in water. They come out and do their job with their heads down. There's nothing more important that's happened to our restoration program than adding bison."

Bison aren't the only keystone species at the refuge. Black-tailed prairie dogs are practically cosmopolitan in the Denver metro. Commerce City, the suburb at the edge of Rocky Mountain Arsenal National Wildlife Refuge, has about sixty thousand residents, just a little more than the number of prairie dogs that Lucas said live on the refuge itself. Prairie dogs are a veritable smorgasbord for predators like coyotes and swift foxes, but even more appetizing to North America's rarest mammal, the black-footed ferret. Black-footed ferrets once cruised prairie dog subways across Colorado. These small, nocturnal members of the weasel family (*Mustelidae*) spend their lives in prairie dog tunnels, dining on the occupants and helping keep populations under control. If you ask a black-footed ferret what's for dinner, 90 percent of the time the answer will be "*prairie dog.*" And in the twentieth century, ferret populations crashed following declines of their primary quarry.

Ferrets do best in large prairie dog towns. As unbroken county-sized prairie dog metropoles became scarce, the black-footed ferret began to disappear, dwindling to only a handful of populations by the mid-twentieth century. When the last known black-footed ferrets died off in South Dakota in the 1970s, despite a concerted effort to find another population in the Great Plains that included "FBI Most Wanted"-style posters in rural post offices, biologists believed the species was extinct. But then in 1981, a blue heeler named Shep killed a small animal at a ranch near the tiny hamlet of Meeteetse, Wyoming. When a taxidermist recognized the species, the US Fish and Wildlife Service converged on the ranch and found 120 more ferrets. The animals were vaccinated and radio-collared, and two years later the population rose to 130, but a dual outbreak of bubonic plague and feline distemper struck, thinning the colony to just eighteen animals by 1987. After that, the remaining ferrets were trapped and used to establish a captive breeding program that has been successful in averting extinction. Today, scientists have cloned at least one ferret and, in 2021, the entire captive population of 250 animals received COVID-19 vaccinations when scientists in Denmark found that the disease affects mink, their close cousins. After successful reintroductions, an additional 250 ferrets live in the wild in eight states and Mexico.

The forty-five thousand prairie dogs of Rocky Mountain Arsenal National Wildlife Refuge were enough to support a few black-footed ferrets. In 2015, twenty-eight were released there, a remarkable testimony to the success of the conservation project. In neighboring Kansas, probably home to more black-footed ferrets than any state in pre-European settlement times, only two sites have qualified for reintroductions, and ferrets have now disappeared from one.

Lucas said their ferrets are doing well. The biggest threats are epidemics, particularly bubonic plague. Plague can destroy entire prairie dog colonies, even those with tens of thousands of animals. The disease is spread by fleas, so biologists at the refuge try to control the blood-sucking insects by inoculating prairie dogs. Using ATVs, they broadcast bright blue peanut-butter flavored pellets that contain fipronil, the insecticide in Frontline, a commercial product used to ward off fleas and ticks in household pets. The problem is not just theoretical. In 2019, a plague outbreak struck the colony and generated a health scare in Commerce City, resulting in the cancellation of overnight camping for a Phish concert in a nearby Dick's Sporting Goods parking lot. The refuge was closed to the public while biologists treated the colony with insecticide. Fortunately, no Phishheads or ferrets were infected.

Lucas told me that the refuge has affected the nearby South Platte River in ways that many people don't realize. He said, "the wetland complexes and a handful of native streams, First Creek, Second Creek, and Third Creek, flow in the direction of the river. All of them are important riparian corridors, and they're part of our management plan. I think from an interesting ecosystem perspective, the refuge is a major storm water feature that I don't think people truly recognize."

This goes back to the catastrophic storms that Chatfield and the other reservoirs were designed to mitigate. Floods continue to strike the Denver metro. One of the worst was in 2013, when record rains inundated the Front Range, causing more than a billion dollars of damage. Lucas said the refuge is like a huge sink for runoff. "If we weren't here, or say, the refuge was like the rest of Commerce City, half of Denver would have been underwater. We had a year's worth of rain fall south of the refuge in two and a half days. There was massive flooding in that area, but the refuge held all that water that came north, which protected those downstream communities."

Today, Rocky Mountain National Arsenal Wildlife Refuge receives about half a million visitors per year. It's used as an environmental classroom for local kids; there is a visitor center and some short walking trails. You can go on "safari" and drive the refuge looking for mule deer and bison. If you get incredibly lucky, you might see a ferret. But the refuge gets less visitation than the parks and wild spaces in the mountains. Denver is a Great Plains city, but many people don't realize it. And that's OK with Lucas. He's focused on a bigger goal, because the primary mission of the refuge is not tourism.

"Grasslands across the planet are the number one most imperiled ecosystem that exists," he said. "The main reason is that it's habitable land. These are places where we can farm and live as humans. Anything we can do to improve, maintain, and preserve grasslands is critical."

He thought for a moment, and said, "This is one of the most important things we can do during our lifetimes."

I asked if that was frustrating for a person who has spent twenty years working on a destroyed prairie. He told me, "No. What I've learned in my career is that once native prairie is gone, don't close the book. It can come back. I think one of the more interesting chapters in conservation is where we are at now in the United States. We have basically protected all the easiest stuff, we've created big national parks, big wildlife areas, big wildernesses, and that's important. We're not finished with that, but most of the portfolio is in place." Lucas continued, "We must preserve that and guard it against encroachment. But if we're going to have any chance at conserving wildlife, we must think about restoration of other important locations. We need to find and restore those locations to create corridors, patchworks of habitat for wildlife. Refuges like the Rocky Mountain Arsenal could be an example of places that can be turned into something that's important for fish, wildlife, rivers, and humans. Don't close the book on grasslands."

Rocky Mountain Arsenal National Wildlife Refuge is at not only a metaphorical crossroads but the literal crossroads of the South Platte. North of the refuge, the river flows into the Colorado Piedmont, a low valley nestled between the foothills and mountains to the west and the high-elevation prairies to the east. From there, the river becomes the lifeblood of the Colorado plains on its journey north and east toward its junction with the North Platte in western Nebraska. I wanted to paddle the meltwater in those plains, especially the river's final miles in Nebraska. This would turn out to be more difficult than I imagined.

The annual melt that once avalanched from the mountains has gone dormant now except in the wettest of years. The South Platte has changed significantly since pre-European settlement times. After leaving Denver, it takes a northern bent through Commerce City before reaching Brighton, and then veers west three times against low foothills before flowing past the mouth of St. Vrain Creek, a stream that drains part of the Colorado Piedmont north of Boulder. From there, the river turns northeast, passing south of Greeley, crossing Highway 34 in agricultural country, then strikes east through the high plains oases of Orchard, Weldona, Fort Morgan, Snyder, Merino, Atwood, and Sterling. From there it enters the little-known Colorado Sandhills just shy of the Nebraska state line. Finally, the river parallels the North Platte, a sibling in name alone, as it makes its own journey from headwaters near Walden, Colorado, via southern Wyoming. The two rivers join to form the Platte River in North Platte, Nebraska.

According to scientific expeditions such as Stephen H. Long's ascent of the river in the summer of 1820, the South Platte was once nearly treeless, broken frequently by islands composed of gravel and sand pushed downstream during the melt. Treeless sections a dozen miles or longer were common, broken by narrow fringes of thunderstruck timber, oases where bald eagles built nests and mountain lions crouched in pursuit of elk that came warily to drink at water's edge by night. In spring and early summer, the river was wide and shallow, then rose to brimful during peak melt on still June nights. After that, it retreated into necklaced pools for the remainder of the year, pools that connected dry sections where the river ran below scorching fields of scrub and sand. Flocks of migrating sandhill cranes and geese swarmed the South Platte valley twice a year during migrations.

The botanist Edwin James chronicled Long's 1820 expedition.

He was struck by the cornucopia of animals the expedition encountered in the treeless prairie. On their first week traveling west from the junction of the South Platte and North Platte, Edwin wrote that they encountered a herd of 10,000 bison that passed them in a single day, "grazing in undisturbed possession, the verdant plain; to the right and left, as far as the eye was permitted to rove, the crowd seemed hardly to diminish."[13]

The river was famously shallow. Riders could cross without dismounting from their horses or mules. The expedition observed bald eagles and throngs of wild horses. James wrote of the horses, "their playfulness, rather than their fears, seemed to be excited by our appearance, and we often saw them, more than a mile distant, leaping and curvetting, involved by a cloud of dust, which they seemed to delight in raising." The group also reported mirages in the hottest parts of the days that produced "the perfect image of a pool of water in every valley."[14]

By the beginning of the twentieth century, the South Platte was already becoming more constrained as meltwater events became less erratic due to new reservoirs and irrigation ditches. Today, it traverses a narrow, rigid channel, with both banks almost continuously girded in cottonwoods from Brighton to North Platte, in part because of subjugated flows and severe droughts that allow seedlings to sprout and take root without the scouring of spring floods. As river geologist Ellen Wohl wrote in her fascinating book, *Wide Rivers Crossed*, in a single human lifetime, the South Platte changed from a "turbid, roiling" stream with channels "a mile wide and an inch deep" to a "narrow, limpid [river] flowing quietly between shaded green banks."[15]

Limpid or not, the river crosses the entirety of eastern Colorado and, at peak snowmelt in a wet year, it should be possible to paddle from Denver to North Platte, and from there, on to the Missouri, the Mississippi, and the Gulf of Mexico. A few stretches are popular with paddlers, including a tricky nine-mile float with mandatory portages between Brighton and Fort Lupton, and a popular run between Evans and Kuner. Not many people, however, take on the river from Denver to Nebraska at one go because of low-head dams, dismal flows, and confusing river-access laws on private property.

A modern adventurer, Rose Shirley, tried it in 2021, paddling her kayak from Confluence Park to North Platte, Nebraska, as part of a ten-year personal expedition across North America by foot and kayak. It became her biggest challenge, and a disaster near Milliken, Colorado, made her rethink why she'd started the journey in the first place.

Shirley seems fond of ten-year plans. Although she spent only six years in the navy, she followed that with ten years driving a truck and ten years working as a cosmetologist cutting hair. But she was languishing. Struck hard by the urge to chuck it all and *migrate*, Shirley ditched her life in Seattle for an adventure she had been planning in her head for years. In 2019, she and her dog Oreo began a new ten-year plan to cross America by foot and kayak, hiking parts of all the national scenic trails and paddling as many rivers as possible that connect to the "Great Loop," the rough oval of waterways that includes the Gulf and Atlantic coasts, the Great Lakes, and the Mississippi River. The tradition of life-altering American walking and paddling pilgrimages is a storied one, but few modern versions include so many miles kayaking seldom-traveled grassland rivers.

Shirley and Oreo began on foot, hiking east from Seattle. After they crossed the Continental Divide and reached Whitehall on the Jefferson River in Montana, she bought a kayak big enough for all her stuff and a few weeks of supplies and dog food. After resting for a week, they paddled the Jefferson to its confluence with the Madison, and then down the Missouri, reaching Helena at the end of the first year. The next spring, they followed the Missouri east, spending at least two days paddling up each tributary they passed, including the Medicine Bow, the Little Sioux, and the Little Missouri. When she discovered how different the tributaries were from the big river, Shirley decided to find a long tributary that she could get to know in a deeper way. She needed to resupply in a big city and had always wanted to spend time in Denver, so she chose the South

Platte. After studying maps and speaking to friends she made on social media, Shirley had a basic idea of how long the trip would take
but couldn't find anybody who had paddled all the way from Denver
to North Platte in a single go.

Luckily, she picked a wet year. When the river rose from the
spring melt following an exceptional winter snowfall, Shirley and
Oreo launched their kayak at Confluence Park and paddled toward
the plains. For the first month, they camped every night on sandy
washes or slept in the boat. She saw deer, bald eagles, skunks, beavers, and heard elk bugling. After its northern turn into the Piedmont, the river starts to flow through private property. Paddling the
South Platte through the Piedmont is challenging because of river
diversion dams. The first one Shirley and Oreo hit was the Fulton
Irrigation Ditch in Henderson, an obstacle that required a full portage. From there, it became a daily pentathlon of portages. Five
dams per day on average. She told me that, on most of the dams,
"there were no warnings or markings. On the very first dam I came
to, I got hung up on the rocks. When I got past that one, I made
it to the next dam and was able to portage. That was the pattern
from then. I'd make about twenty miles a day, portaging four or five
times, then sleeping on the prairie."[16]

Most landowners she met gave her tips on sections of the river
they had paddled and wished her luck on her journey. "Most people
were really nice, and a few were helpful." A farmer near Brighton
was an exception. "He aimed a shotgun and threatened to shoot me
from across the river. I was portaging a dam and had my hands full.
He threatened to call the sheriff. I told him that would be fine, I'd
meet them at the next bridge, but when we pulled up, nobody was
waiting for us."

The dams continued to be a daily source of mental fatigue. She
said, "It was spring, and I had the water from the melt under me.
The biggest problem with floating the river are the dams themselves, and the riverbank modifications prior to the dams. At one
place, there is a mile of crushed cars before the dam, so it's almost
impossible to get off the water because they're stacked up so high

and steep you couldn't climb the bank. I eventually got over the edge, so I didn't plunge over the dam, but it was one of the hardest things I've ever done."

Shirley and Oreo continued for weeks, making steady progress, camping beneath the cottonwoods and living off food she'd carefully partitioned from the big resupply in Denver. On Day 36 out from Confluence Park, she passed Milliken, a town of eight thousand that once was a hub of sugar beet production. Paddlers can recognize low dams on small rivers from various clues. Often the water deepens and slows, and the line of trees on either side drops abruptly. At Milliken, Shirley said, "I didn't see the dam, I missed all the signs. I hadn't noticed it on any of the maps, so I began the day paddling, not thinking I'd reach the first dam until afternoon. But it came almost immediately after we pushed off. By then it was too late. I paddled to get us to the left, or to the bank and even looked for a notch or some concrete we could use to grab onto, but it was too late. We crashed off the dam. At first, I swung around and almost made it, but then it flipped my kayak, and we tumbled over into the river."

Low-head dams that span rivers are deadly, creating hydraulic sieves that can be impossible to swim out of. Shirley herself was lucky—she was on the end of the dam and the current pushed her back into the river. She lost her electronics and all her supplies. But when she retrieved her boat, she swam downstream and found that Oreo hadn't survived the accident. It's still hard for her to talk about. "I took Oreo all over the country with me, and I know he really enjoyed his life. He was a happy-go-lucky dog."

Shirley sat under the Colorado Highway 60 bridge for the next two weeks, devastated. "I stared into space because Oreo was lost, I didn't have my phone, and I was cut off from my support system, the paddler community. I just sat there, eating through my food. Mourning. I didn't know whether I would continue. It wasn't the first time I'd gone through something hard like that, but it was a bad time."

Eventually, she was able to replace her phone and get a new

paddle. She found a place to bury Oreo near the banks of the river. By the time she resupplied and was ready to continue, there were new concerns. "During the two weeks I spent there, I watched the meltwater pass. The river went from ten feet to two. That's when I started spending long parts of each day dragging the boat through places with six inches of water. That's what you've got to do to keep going. For about a mile or two before the dams there would be good water, but the seasonal pulse from the melt was over."

The dams got farther apart in the Colorado Sandhills, but she had enough water to reach the confluence of the North and South Platte Rivers before she lost the spring melt completely to irrigation and diversion. Although she didn't continue along the Platte River to Omaha, she had paddled the entire grassland section of the South Platte River, living free on the plains, and riding the spring melt.

The South Platte and North Platte Rivers complete their journeys across the plains in North Platte, Nebraska, a small town that is officially the birthplace of the Platte River proper. Like its southern sibling, melting snowpack fuels the North Platte, a grassland river that originates in the Rabbit Ears Range near Rocky Mountain National Park. As it descends from the Continental Divide, the North Platte sweeps up into Wyoming before threading across 700 miles of grasslands. The river is dammed near Ogallala, Nebraska, to form McConaughy Lake. Legal compacts preserve much of the North Platte's water for Nebraska, enough to protect cranes and herons on the Platte and solidify the state's reputation as the cornhusker state.

The valleys of the two rivers join below the hydroelectric power plant at Kingsley Dam. In their final miles between Hershey and the confluence, they perform their own crane-like dance for forty miles, at some points less than four miles apart. For most of the year, the South Platte runs a few inches deep, or under the sand itself, while the North Platte, its current spared for irrigation further downstream, runs deeper and is easier to paddle. The North Platte

River Water Trail between Hershey and Buffalo Bill State Recreation Area is a popular paddling route.

I traveled to Hershey in June to paddle the South Platte, along the final fifteen miles before the confluence, a stretch that could become a popular paddle destination if stream flows were more dependable. To be accurate, this was my fourth trip to Hershey in search of the river. I had timed my trips the previous three Junes to ride the melt, but there wasn't enough river in any of the years to put my boat in the water. Each year, I paddled different Colorado stretches, only to be defeated by low flows by the time I reached Nebraska, although I did consolation floats on the North Platte River Water Trail. But on this June 16, the morning of the strawberry moon and the historical peak date for the spring snowmelt, I got lucky. The previous week it had rained hard each night across eastern Colorado. Thunderstorms recharged some of what the irrigation flumes had diverted, promising to give me about two feet of good water, enough to guide my canoe without too much dragging. For the last few days, Chris had shuttled me between sections of the South Platte northeast of Denver to take advantage of the favorable flows.

On the last day, he helped me pack the boat, then drove my car to the takeout before he went back home to Boulder. I was on my own for the final stretch. Navigating the sandy braids of the river, I thought about the North Platte four miles away, deep enough that paddling was optional. From the start, I had to pick my way carefully downriver. Reading the current was easy most of the way because the stream split into a main channel and several shallow channels a few inches deep. The trick was choosing the darker water of the deep braid. Compared to the Kaw and the Grand, tallgrass rivers both, the banks of the South Platte were low, stabilized by cottonwood forests on both sides. I wondered how different the river would have looked back when the prairies extended down to the banks. Even with the trees, the South Platte was easily the most braided prairie stream I had ever paddled. Though it was already ninety degrees in late morning, the east wind gave a playground push and felt cool against my cheeks. The small riddles to solve to stay in current focused me on the task. I relaxed into the pursuit of "deep enough," knowing that, in just days, the hot sands would drink the water to a trickle.

Interstate 80 should have been close enough to hear cars zooming by, but the valley absorbed the clatter of traffic. The sandhill cranes had finished moving through six weeks earlier. Smaller birds swarmed the river now. Redwing blackbirds paraded by with crimson shoulder patches and families of western kingbirds lined up on wiry cottonwood branches, chattering to each other with voices like rubber ducks. Wood ducks and mallards led broods of fluffy ducklings into eddies to hide. My companions on all river journeys, great blue herons, did their best to ignore me at every bend. Diversion ditches threatened to lead me away in places where canals bubbled out into the fields. I had to be mindful to stay in the main river and avoid the diversions that flowed out into canals.

There was no rushing this float; even in good water I was making less than two miles per hour. The mental game to keep finding channels was more exhausting than the physical paddling. I thought how frustrating it must have been for Rose Shirley as she and Oreo alternated between paddling and dragging the kayak. I stopped at a sandbar next to a tea-colored sluice of reeds growing below a stand of Russian olives, themselves alien competitors— Colorado Doctrine be damned—for the little unallocated water that made it this far. What looked like a small wooden statue on the edge of a sandbar didn't move as I neared, but when a large bald eagle made a close flyby, I realized the statue was a resolute adolescent eagle. Not old enough yet to have earned its white feathers, the junior bird didn't fly off, in obvious defiance of its mother, not budging when I paddled by. The mama bird flew back to her cottonwood branch to wait. Canada geese covered a glistening meadow in fields beyond the trees.

Compared to other grassland rivers, the South Platte and its sandbars were remarkably clean, like a trash crew had just been through; that is, if you didn't count the invasive salad of Canadian thistle, Queen Anne's lace, and the dried brown stalks of great mullein, the soft-eared herb that has spread across the prairies after being introduced from Europe in colonial days.

The water didn't hold out to the end. The last miles were more dragging than paddling until I reached my takeout. I hadn't paddled all the way from Denver, or from the mountains, but I had ridden the meltwater on this vital artery that has sustained the people and animals of the arid plains for millennia. Unlike Mark Twain, I didn't find a melancholy river, only a melancholy legacy of environmental abuse that is beginning to change.

Grassland river revival is alive on the South Platte. Thousands of people work and play every day along its banks, and millions of dollars are spent to help make it a centerpiece of contemporary Denver. But the river is still beset with challenges driven by population growth in the Front Range. Though cleaner than it once was, the river is still polluted. In 2013, Denver Mayor Michael Hancock set a goal to make the river safe enough to swim and fish in by the end of the decade. *E. coli* levels, however, haven't changed much since then, and, despite major improvements to wastewater utilities, the river still carries enough toxins to cast doubt on whether this will ever be attainable, especially considering the growing population.

Farther east in the plains, groundwater pumping threatens biodiversity in the warm water section of the river below Denver. While the water table has not yet significantly declined, the basin immediately to the southeast of the South Platte could serve as a warning. From 1949 to 2005, irrigation from groundwater pumping in the Arikaree River drainage of eastern Colorado increased from 8,500 square kilometers to 63,000 square kilometers. Groundwater extraction from the aquifer exceeds recharge, resulting in a "dewatering" effect on area streams and rivers. Mean surface stream flows decreased by 60 percent during this period. As streams shrink, the fish communities they support become more homogeneous. Unlike in the Purgatoire, where native fishes continue to thrive, the fish communities in the Arikaree have shifted from predominantly large-stream species to small-stream species—a significant change that could have ripple effects throughout the riverine ecosystem.[17]

Another risk is tougher to quantify. The Great Plains are warming. Droughts could become longer and more frequent. On my way to ride the meltwater, I drove through the Colorado Sandhills in the extreme northeast part of the state, a region many people even in Colorado are unfamiliar with. South of Ovid, row crops cling to the river for about a mile, and then fine sage prairies of bunchgrass rise into the hills. Playas and buffalo wallows dot green pastures draped over sand dunes evident in road cuts. Ovid gets about fifteen inches of rain per year, twelve inches more than the Sahara Desert. The dune fields are an archaeological reminder of frequent severe droughts in the past. They have been "active" in recent geological time, between 650 and 1,000 years ago. Geologists believe that reactivation of the dunes in the future seems likely, especially with increasing megadroughts from climate change. That could dry up the South Platte for years at a time as temperatures rise.

After my float, I drove across the town of North Platte and spread out a tarp at Cody Park to eat beans and tortillas and rehydrate from the hot day. Both the South Platte and North Platte Rivers bring waters from Colorado to the edge of Nebraska. The summer before, I'd rented a kayak from Dusty Stables outfitters to float the scenic section of the North Platte. When the proprietor, Dusty Barner, dropped me at the put-in near Hershey, we hung out for a while and watched the river surge from under a low road you'd have to duck or portage if the river was any higher. The current bubbled through a stand of reeds seething with redwing blackbirds. Barner said, "I don't think people outside Nebraska realize the rivers we have here. The North Platte. The South Platte. The Platte River itself that everybody knows. But within a hundred miles . . . The Dismal. The North Loup. The Middle Loup. The Calamus. The Sandhill. The Niobrara."[18] He stopped talking, continuing to look at the river. After a while I said aloud what I thought he might be thinking. "Almost too many rivers for one lifetime."

A kingdom of grassland rivers, all bound for the Missouri. The name "Nebraska" means, in the language of the Omaha people, "flat water." I wanted to paddle them all. One, however, had been ringing in my mind for months: the Niobrara, a river that many believe combines the best of Nebraska rivers and biomes in one great stretch. The Niobrara is haunted, singular, difficult to describe; filled with fossil flows that spill out as waterfalls from punctures in the Ogallala. Strange forests of the past grow along its banks. A riverine seam, it gathers almost every kind of prairie found in North America. The Niobrara gets more recreational use than any grassland river in the country. I know people who moved to Nebraska with no job or prospect of getting a job just to be near the river. It is on everybody's bucket list, and if I didn't despise the idea of bucket lists themselves, it would be at the top of mine as well. When I saw it for the first time, I never thought about Nebraska the same way again.

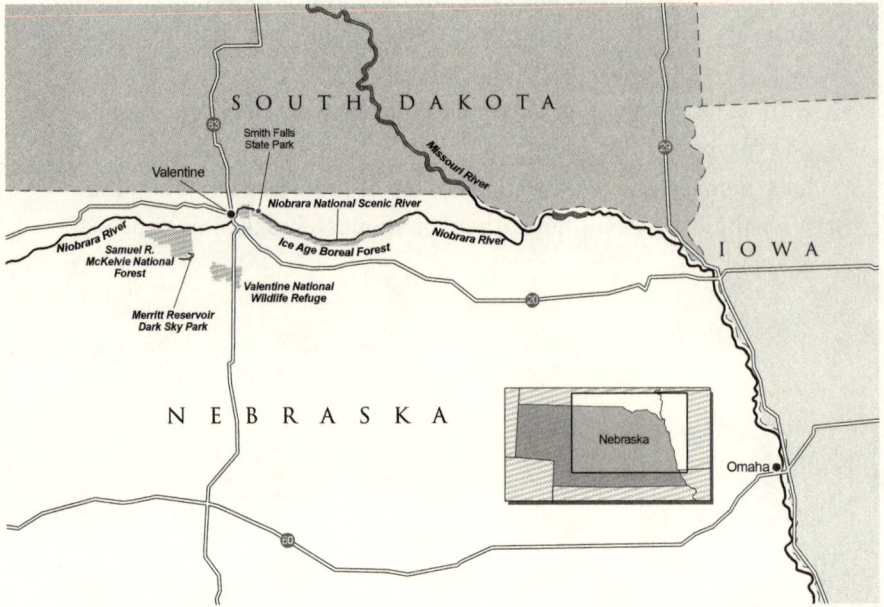

MAP 7 The Niobrara River

CHAPTER 7

NIOBRARA

It was the season of the yellow puccoon. Everywhere, the prairie reached up with bouquets of creamy *Lithospermum caroliniense* flowers arranged atop furry stalks of blithe greenery. It was always the season of wind. In the Nebraska Sandhills, wind and time flow together in syncopated tandem, their invisible orbits entwined in a never-ending dance.

On the edge of a spoon-shaped plateau perched a breathtaking four hundred feet above the Niobrara River, the wind roared, gusting forty miles per hour. Only the cinched chin strap kept my hat from blowing off to the Rosebud Reservation ten miles across the South Dakota line. The overlook, in a grove of ponderosa pine perched atop the precipitous drop-off, revealed a shocking panorama.

Below an oak-covered ridge, the river marched with military precision for a quarter mile before hard-pivoting left at a bend where a spring poured down over the face of a beautiful sandstone cliff. In the valley of the spring, oaks, elms, and linden trees, typical of an eastern deciduous forest, dissolved into an altogether different environment—one of the strangest woodlands in North America, a relic from a lost age. Beyond the thread of river forest, a scroll of native grasslands—northern mixed-grass prairie, tallgrass prairie, and, unique to this landscape, Sandhills prairie—unrolled toward the horizon in the four cardinal directions. Like the wind and the river, the entire landscape was *moving*—waving and rolling. Six different plant communities were visible from the overlook. The spirit of the river somehow unified them all. The Nature Conservancy calls the seventy-six-mile Niobrara Wild and Scenic River the biological crossroads of the Great Plains. A little-known national treasure, the river flows through one of the wildest grassland corridors left in America.

In the 2020 census, Nebraska had less than two million residents—fewer people than in the thirty-five largest US metropolitan areas.[1] But the state is delirious with rivers. None can eclipse the Platte, the great crane river; or the Missouri, which emerges from Gavins Point Dam—the last mainstem dam before the big river laces up for a final channelized run to the Mississippi—to receive the other rivers after they've carved their way through layers of geologic time on their journey to the Gulf of Mexico. Tallgrass rivers like the Big Nemaha, the Big Blue, and the Elkhorn course through the mostly plowed eastern third of the state. The real action, however, is in the Sandhills. The Dismal, the South Loup, the Middle Loup, the North Loup, the Cedar, and the Calamus Rivers are among the truest of prairie rivers in the country. Wide, shallow, and braided, these rivers flow year-round, strong and steady, winding through broad grassy flats, feeding on groundwater from the healthiest remaining quadrant of the Ogallala Aquifer.

The Nebraska Sandhills comprise the largest region of dune formations in the western hemisphere. Spanning 20,000 square miles—a quarter of the state—the Sandhills harbor extensive never-plowed grasslands, wild rivers, tens of thousands of namesake sandhill cranes, uncountable natural lakes, wet meadows and marshes, and a fascinating flora and fauna adapted to the harsh environment. Despite its ecological significance, the Sandhills attract few visitors compared to other notable ecoregions, offering a rare opportunity to feel lost in the middle of the endless prairie.

Formed during the late Pleistocene from sands pestled out of the Rocky Mountains by glaciers, centuries of wind blew the sediments east, creating great dunes. Beginning along the South Platte in northeastern Colorado, the dunes achieved their zenith in western and central Nebraska. The immense formations—the tallest reaching four hundred feet—act like a giant sponge, trapping most of the seasonal rainfall. This constantly replenishes the Ogallala Aquifer and hoists the water table to form lakes and wetlands. Nutrient-poor soils rendered the ample water irrelevant for agriculture and spared the prairies. Instead, a livestock grazing culture evolved, and today, cattle outnumber people.

Aquifer-charged groundwater propels rivers like the Dismal, a seventy-mile stream where water boils to the surface above punctures in bedrock. In interdune depressions, the water table rises high enough to form thousands of shallow marshy lakes—wetland hotels for waterfowl during their vernal and autumnal migrations.

The Niobrara River nicks the northernmost flume of the Sandhills on the river's 568-mile journey from the shortgrass prairies of Wyoming to the Missouri River. Translated from the Omaha-Ponca term that means roughly "broad water" or "water spread out," the liminal river not only joins six different ecosystems but also serves as a biological ecotone: here dozens of species reach either their eastern or western limits, and sometimes they hybridize to create unique organisms found nowhere else.

I'd always wanted to paddle the section of the river christened "wild and scenic" by the National Park Service in 1991. In grad school, a geographer friend told me that spring-chilled microclimates allowed stranded remnants of a long-vanished Ice Age forest to freakishly survive into modernity. I also wanted to explore the sandhill prairies protected in reserves such as Valentine National Wildlife Refuge; and I wanted to explore the "national forests" of planted trees that strangely also contain large grasslands. One 4,000-acre parcel of prairie and forest in Fort Niobrara National Wildlife Refuge was in such primordial condition that Congress designated it a wilderness. Along the thousands of miles of grassland rivers, the Niobrara was the only one I had ever heard of flowing through a congressionally sanctioned prairie wilderness.

Running a thin-skinned Kevlar canoe was a bad idea.

Chloe and I had arrived at the upper Niobrara National Scenic River Trail with a few of the Kaw River guides and the used Wenonah canoe I'd acquired the previous winter. Our plan: paddle all morning, hike into the riparian zone with one of America's foremost Ice Age forest experts in the afternoon, then paddle to Rocky Ford Rapid before sundown—a long day on the river. As soon as we launched below Cornell Dam (where we had each paid a one-dollar launch fee for the privilege—what else as valuable as access to a wild and scenic river costs only one dollar?), we scraped the delicate hull of the Wenonah across the jagged top of a moss-covered rock. A sloppy start.

Since it was constructed at the confluence of Minnechaduza Creek, Cornell Dam has trapped one hundred years of sand. Although after years of siltation sediments now flow over the top of the dam, brawny currents powered by the dramatic hydraulics of spring-powered waters make long stretches of the Niobrara a crane's dance of clear water and rocks. As navigator in the bow, Chloe's job was to guide us around boulders, through small rapids, and to warn of shallow outcrops of the Valentine Formation. At seven miles per hour, the obstacles came at us fast. It was a tough job.

Ripe plums grew in briars along the banks, so we pulled out below the first bend to check the canoe for damage and gorge on what we could knock down with our paddles. Wild plums known by many names occur throughout the southeastern United States, but the most common variety in the northern Great Plains is Sandhill Plum, *Prunus angustifolia*. The bushes sagged with small wine-colored fruits, each with a worm we had to flick out before eating. Chloe said she felt rich.

We had slept the night before at Sharp's Campground, one of many outfitters that run tubing and kayak trips, rent gear, and provide campsites to the tens of thousands of visitors each year. Two screech owls on opposite sides of the river had sung us to sleep with their strange butterscotch tremolo. I thought the calls were sweet nothings between mates, but Chloe said they more likely meant *"You come to my tree, I slit your throat."* Turns out the owls use the monotonic trills to stay in contact with their families, not to assert territory. They have a horse-like "whinny" call for that.

We shuttled the boats with our own vehicles, so we were able to put in several miles above Smith Falls State Park, the common launch point for day trips. That meant we had the river to ourselves before joining the big party of tubers later in the afternoon. In no hurry, we ferried across the river and beached the boats to see Fort Falls. Chloe and I hiked the steep trail to the top of the plateau and looked out across the river valley. "Ye gods," she said, "the land that time forgot." I laughed because I'd never heard her say "ye gods" before.

Back in the boats, a series of narrow canyons squeezed the river into small class I and II rapids. Nothing more than freckles in the current, we nevertheless hopped out onto a gravel bar to scout the trickiest set. It had clear channels to follow between boulders but required a left side-maneuver in the tightest spot. Chloe pried us through it skillfully. As I anchored with my paddle to turn, however, the starboard side of the boat swung around and scraped across a hidden rock and opened another laceration on the Wenonah. I winced when I heard the squeal of Kevlar, but when we spilled out into the eddy at the end of the rapid, the soothing ambience of spring water seeping down a yellow cliff restored the serenity of the canyon. If we were going to peel the carapace off the new canoe, we would do it in the presence of sublime springs.

The first miles of river passed a series of canyons and at least a dozen separate springs, although I lost count. Springs ran from forests on both sides of us. The river meandered at will, untouched by channelizers who had had their way with most of our grassland rivers. When it got hot, the river would enter another wooded canyon that cooled things off.

We pulled into a slackwater below the last of a short series of rapids, and the others waded into the water. Vireos and warblers flitted about in an oak that branched over the river, the fall migration already in swing. A butterfly landed on Chloe's shoulder, and when she slowly turned her head to look, it moved to her paddle. Two white-tailed deer shied from shadows to the water's edge, bothered more by flies than by Chloe and me sitting as still as possible to watch them drink.

FIGURE 7.1 Smith Falls surrounded by Ice Age boreal forest at Niobrara Wild and Scenic River. Photo by the author.

She noticed movement downstream beyond the deer and got out her binoculars. "What are those fat things?"

I took the field glasses. A pickup truck was backing around slowly on a flat, angling to get a long trailer down close to the water. Three guys began unloading inflated floating devices off the trailer. An old school bus had pulled up to the launch and people were getting out with their coolers and pool toys and heading down toward inflated tubes molded into the forms of floating chairs, loveseats, porch couches, full-on sunken living rooms, and classic canvas-hided black rubber donuts, each fitted with drink holders aplenty. We paddled toward them, and the river swung into another secluded meander and then straightened out as a light mist began to fall.

We had reached our stopping point for the morning, Smith Falls State Park, home to Nebraska's most spectacular waterfall. From the park down to where it leaves the Sandhills, the Niobrara runs through a series of light rapids and tight turns, turbocharged by the effluence of cool springs, making it the most popular stretch of river and a bustling scene of activity during summer. But we had stopped here because the park is also a window into a lost time, where the last Ice Age never completely ended.

The cabin air was incense from a day-old delivery of split cedar logs, now stacked in two neat rows along the floorboards of Smith Falls State Park headquarters. Superintendent Amy Kucera offered us coffee and told us the cedar came from a woodlot a few miles downstream on the Niobrara River. The logs were destined for the campfires of paddlers and day campers, although cedar wood burns hot and requires vigilance in dry environments because of its tendency to spew embers.

Wildlife biologist Gordon Warrick's light green boonie hat went on the table and his round wire-rimmed eyeglasses on his lap as he wiped them with a bandanna. Warrick was two years retired from the National Park Service, where he had worked on natural resources management for the Niobrara National Scenic River. He had created an aspen management plan, not an ordinary task for somebody managing natural resources at prairie rivers in the Great Plains, but the forests of the central Niobrara valley are anything but ordinary.

"Eastern red cedar?" I asked Kucera, nodding toward the firewood. Warrick and Kucera had fought the prairie-threatening tree their entire careers. Kucera said, "We can't sell enough cedar firewood to qualify it as a proper prairie management technique, but it does smell nice."[2] Anything that encouraged efforts to thwart the encroaching green glacier of eastern red cedar seemed like a good thing. Although a native species, the tree, a member of the juniper family, functions as an invasive in prairies that aren't burned or grazed at regular intervals.

Smith Falls State Park is one of the best places to explore the most unusual plant community of the Great Plains, the northern boreal forest. The misfit biome occurs on the south side of the river in steep spring-branch canyons that extend in places only a quarter mile before Sandhills prairie replaces them. Intimate micro-climates refrigerated by the waters of springs and waterfalls that tap deep into the heart of the Ogallala, spring-branch canyons create slivers of habitat where the magical forests of birch and aspen can survive.

On a map of ecosystems, northern boreal forest—also called Ice Age forest, snow forest, or taiga—appears as a green ribbon wrapped around the earth immediately below the southernmost arctic tundra. It represents a quarter of the world's forests and is relatively uninhabited by people. During the Ice Age, when the climate was cooler, these forests covered much of the future northern United States, but after the glaciers retreated 10,000 years ago and the climate shifted to a warmer and drier period, boreal forests retreated along with the ice. Most of them, at least.

The geology of the Sandhills and their unique interaction with the Niobrara created a refuge that allowed the remarkable ghost ecosystem to persist to the present. In geology lingo, a "formation" is a fundamental unit of the earth's strata. Geologists map formations to better understand the geology of a region and the systems that depend on it: hydrology, soil chemistry, plant communities and the like. Geological formations usually have hyperlocal names. Rain that falls on the Sandhills either runs off toward the river or soaks into the Valentine Formation, the porous silt and sandstone layer beneath the dunes. The Valentine Formation is named for nearby Valentine, Nebraska. The Rosebud Formation—named for the Rosebud Reservation of the Rosebud Sioux, a band of the Lakota people—underlies the porous water-bearing Valentine Formation and acts like a concrete storm drainage culvert. When water seeping through the Valentine Formation hits the Rosebud Formation, it planes out toward the Niobrara, spilling into the river and surrounding draws from dramatic springs and waterfalls. The artesian waterworks are thrilling, and the cooling effect fosters a shallow but stable microclimate where the boreal forest manages to persist, hundreds of miles from its nearest counterpart and a thousand miles from the heart of the Canadian taiga.

We left park headquarters and walked down to the footbridge

over the river. A Nebraska native, Warrick had been coming to the spot for almost his entire life. While attending college in Kearney, he worked summers at the Nature Conservancy's nearby Niobrara Valley Preserve and ran canoes for an outfitter. We crossed the bridge and followed the path leading to Smith Falls, the highest waterfall in Nebraska. Two hundred acres of the state park, including the waterfall, is leased from a private owner. The park, along with an additional thirty-four acres owned by the state, provides public access to the wild and scenic section of the river.

As soon as we reached the trees below the waterfall, the temperature dropped. Warrick explained how this was unusual in the Great Plains. "We've got the bluff line on the south side of the river that is very steep because the river is downcutting. That creates a sharp north-facing slope that doesn't get a lot of sun, it doesn't get the desiccating south wind. It's in the shade much of the time. So that's a degree of microclimate right there. But then, going into a spring-branch canyon, the evaporative effect further cools things and we get even more trees. This creates a virtually closed canopy, and a cool, moist climate."[3]

Beyond the thermocline, we began to see birch and aspen, two keystone species of the boreal forest. It was astonishing that this waterfall and the thin, refrigerated wafer of woodland had remained stable enough to support this oddball forest since the last glaciation. Similar spring-branch canyons south of the river harbor Ice Age forests along a forty-mile stretch.

We stopped at a cluster of the smooth-barked aspen, their leaves trembling in the wind. I initially mistook the trees for quaking aspen, the familiar tree of the Rockies. However, true to the crossroads nature of the Niobrara, these aspens were an east-west hybrid. *Populus grandidentata*, called big-tooth aspen or white poplar, is an eastern species. *Populus tremuloides*, or quaking aspen, is a western species. The Niobrara aspens are a hybrid of the two. Warrick said, "During the ice age, both species occurred here. According to pollen records, we know big-tooth aspen hasn't been here for at least 5,000 years."

That means the western and eastern species crossed their genes at least 5,000 years ago. The pollinizing event produced hybrid seeds and a few sprouted into trees and grew to maturity—precisely three, according to Warrick. "In the 5,000 years since the big-tooth aspen disappeared, the hybrids survived by vegetative reproduction." With vegetative reproduction (as opposed to sexual reproduction where pollens fertilize female flower structures to create seeds), the hybrid aspens have miles-long underground networks of roots that send up shoots of new aspens at regular intervals. Warrick added, "There are three distinct clones in the middle Niobrara valley, two males, and one female."

For forty miles, the aspens, whose branches reached out to one another to form the crown of the overstory, were just three individual organisms. At least, they started out that way. Some of the forest has since calved island colonies of the clones from the main root systems of the parent trees. Warrick said this extremely limited gene pool presents a risk for the long-term survival of the forest.

White birch, the other Ice Age tree, is represented by a single species, *Betula papyrifera*. It reproduces both vegetatively and sexually and is somewhat less threatened than the aspens. The big birch with their white papery bark were the tallest trees of the forest. A blanket of sassafras shrubs grew beneath the aspen and birch. Sassafras, like the oaks and linden, is another eastern plant reaching its western limit along the Niobrara.

We lowered our voices at the falls. The hydraulic thundered over the chin of sandy soil like a glistening beard, creating a stain of moss-covered bedrock on the vertical cliff that reminded me of the Shroud of Turin. After emerging from the scour pool at the bottom of the waterfall, the discharge cascaded down the hill and dumped into the river near the footbridge. A typical prairie river might be fed by one or two half-hearted springs per mile, but Smith Falls spring, though spectacular on its own, was one of hundreds in tight proximity, watering the soul of the river and its wonderous forest. I'd never seen anything like it.

Below the waterfall, Warrick took a thin side trail paralleling the river up into the bluffs to explore more spring-branch canyons. Raspberry and rose vines kept grabbing my jeans as we hiked. Red columbines bloomed beside poison ivy shoots. Soon, we were a hundred and fifty feet above the river. As the din of the falls receded, we began to hear birdsong again. Warrick identified a wood pewee, a small flycatcher with eastern and western varieties. Both occur along the river but are difficult to distinguish from one another except by their calls.

The intermingling of species in the central Niobrara River valley is not limited to trees. When a flicker flew over us, I asked if it was the yellow- or red-shafted variety of the jaunty speckle-chested woodpecker. Warrick thought it could be a hybrid. Because the ranges of several closely related eastern and western species overlap, the forest is a hot zone for avian hybridization. The yellow- and red-shafted varieties of northern flicker, eastern and spotted towhee, Baltimore and Bullock's oriole, indigo and lazuli bunting, rose-breasted and black-headed grosbeak, and eastern and mountain bluebird are all thought to hybridize to some extent in the Niobrara woodlands.

The trail was steep near the top of the canyon. At the summit, a few ponderosa pines—the keystone species of the western coniferous forest—grew among their deciduous and boreal brothers and sisters. Below us, a small stand of fledgling aspen was fenced off to deter deer and beaver. At the top of the ridge, it was easy to see that many of the big aspens and birch on the slope above the young trees were dead. Tree mortality and standing deadfalls are expected in old-growth forests, but with so little genetic diversity, biologists keep close watch on the Ice Age trees, especially the aspen.

Warrick spoke about one of the giant, dead standing birches like a personal defeat. "Two years ago, we ran a fire through here to burn away the understory. The fire top-killed some of the big trees, not all of them, but ones like this." He pointed to a steep ravine and a grove of toddler birches. "That's the big dead tree trying to regenerate." At least fifty feet from the dead trunk, the tree's subterranean component had reincarnated through vegetative propagation after the death of the "bole," as Warrick called it. For many trees, life does not end when the trunk dies. "The trees are all competing with one another for sunlight and nutrients," Warrick said. "They are all interconnected by this root system. After something like a fire, any-thing that sets them back to the ground, this stimulates a flurry of vegetative growth in the short term."

Warrick inventoried the aspen and birch on the hillside as we walked back down to the river. He seemed to know each tree per-sonally. "Historically, fire would have kept things in order, but for a long time all fires were suppressed. Today we can't get a hot enough burn to actually kill a lot of the overstory trees that compete with the boreal forest," he said. "I'm not sure how much good we're doing now. The fires will top-kill a lot of these trees, including the aspen. They'll try to regenerate. Like what we're looking at, we have a lot of small birch and aspen. A few of them might reach maturity." As a temporary solution, Warrick said, volunteers clear the underbrush around mature trees to give them a better chance.

Like the floodplains of other near-pristine grassland rivers, the floodplain of the Niobrara was connected directly to the channel. Warrick stopped riverside to show me downy gentian, another na-tive prairie forb with blue flowers. I knew the plant from the Gyp-sum Hills back in Kansas, the landscape that most reminded me of the Sandhills. Warrick said that, historically, tallgrass prairie would have grown between the forest and the river. However, most of the bottoms had been replanted with Kentucky bluegrass. The gentian was a throwback to the old ways.

We weren't alone in the forest. Some of the tubers were using smartphone apps to key out woodland flowers and a guided group was getting ready to cross the bridge near the waterfall. Interest in the ancient forest, Smith Falls, and the river has created a unique opportunity for tourism in this remote part of the plains. Measured by proximity to large cities, the Niobrara is about as remote as any grassland river. Along its 568-mile course, the largest town is Valen-tine, with about 2,600 residents. Cattle production is the predom-inant industry in the region. Most ranches are big and spread out and don't require large labor forces to operate, one of the reasons major cities never developed in the area.

Adventure travel and ecotourism are becoming increasingly significant to Sandhills economies. According to a National Park

Service report, 90,000 people visited the Niobrara National Scenic River in 2022 and spent an estimated $8 million, resulting in an $9.3 million net economic impact in local communities.[4] This was a sharp rise from a net $2.6 million ten years earlier. The broader impact, when considering the entire Sandhills region and activities such as guided hunting, "Wild West" trail rides at local ranches, and biking and through-hiking on the 195-mile Cowboy Trail, a rails-to-trails route that follows the Niobrara and Elkhorn Rivers—is certainly higher.

According to Jenna Bartja, adventure travel specialist for the Nebraska Tourism Commission, the central Niobrara Valley, though still considered a "best-kept secret," is gaining traction with outdoor enthusiasts. Besides paddlers who tube and kayak the National Scenic River, the Nebraska Star Party, an annual gathering of professional and amateur astronomers, has sparked interest in "dark sky tourism." Merritt Reservoir on the southeast side of Samuel McKelvie National Forest is Nebraska's first designated International Dark Sky Park. Bartja told me, "With the National Scenic River and the Dark Sky Park so close to one another, this spot in the middle of the Sandhills, from an adventure travel perspective, is the ultimate escapism. You can be off the grid in a place that is largely untouched by human impact."[5]

As ecotourism draws visitors to the Sandhills and more people explore the Niobrara, concerns about potential environmental impacts have arisen. However, a recent assessment of the river's aquatic ecosystem in the context of tourism indicates that the river remains resilient. Jessica Corman, associate professor of limnology at the University of Nebraska-Lincoln's School of Natural Resources and one of the study's principal researchers, emphasized that tubing and kayaking are a "great way to get people to experience the outdoors and have a sense of pride and place in Nebraska rivers, viewing them as not just conduits of runoff, but an important, integral part of the landscape."[6] Given the Niobrara's status as one of the last nearly pristine grassland rivers in America, and because her research focuses on river systems, she wanted to find out if the goals of recreation and conservation were at odds with one another.

Corman sought a metric that went beyond water quality and chemistry, something that could holistically gauge the health of the river and its immediate surroundings. In its wild and scenic section near Valentine, the Niobrara is a Sandhills river. Corman told me, "These rivers by their very nature have more exposed canopies, because for the most part they don't flow through dense galleries of forest. They also have sandy bottoms. So many rivers in other parts of the country have bottoms composed of rock, bedrock, large cobble, versus our grassland rivers where the bottoms, even in the headwaters, are more sandy. That leads to interesting questions when you have a substrate that is unstable, how does that influence what can grow there in terms of algae and insects. What does that mean for the entire food web of the river?"

Her colleague, David Manning, assistant professor of biology and ecosystem ecologist at the University of Nebraska, emphasized the importance of aquatic macroinvertebrates in assessing the health of rivers like the Niobrara. "Insects, in particular, are really important," he told me. "If we sample aquatic insects, we expect them to respond differently depending on what taxa we're focused on. Different species have different stressors. If something stressful is going on in the environment, there are certain families of insects we just won't find. On the other hand, if the ecosystem is functioning properly, we expect to find a strong diversity of aquatic insects. They're a good bioindicator."[7] Corman and Manning were adept at sampling aquatic macroinvertebrates, and comparing populations with those from other rivers and less frequented stretches of the Niobrara could yield valuable insights.

Manning's graduate student, Jennifer Dailey, spearheaded the study, conducting extensive sampling of the river over a three-month period in the summer of 2020, identifying 4,971 macroinvertebrates from fifty different families.[8] Two of her discoveries were noteworthy. The Niobrara is rich in mayflies from the Baetidae and Isonychiidae families, known for their intolerance for pollution. However, what surprised her most was the discovery of stoneflies. "Stoneflies are something you don't commonly see. They are very, very pollution intolerant and I found quite a few stoneflies in the sections of the Niobrara I sampled. Personally, I'd never seen them in any river."[9] Manning said, "Based on what Jennifer found, and differential comparisons along the river, we believe the recreational stretch of the river is in good shape in terms of its biodiversity from the aquatic insect, macroinvertebrate side of things."

Given the limited number of macroinvertebrate studies conducted on grassland rivers, Corman, Manning, and Dailey's work is not only a hopeful sign but also establishes a crucial baseline for understanding prairie river ecosystems. Their findings suggest that low-impact human activities have minimal consequences for riverine health. This underscores the importance of ongoing protection of rivers under the Wild and Scenic Rivers Act and supports the role of ecotourism in preserving river ecosystems. Such efforts can enhance public awareness and foster a deeper awareness and respect for all grassland rivers.

Later in the afternoon, we waited in line for the ramp at Smith Falls, joining the fifty or more tubes on the water ahead of us—the first traffic jam I'd ever experienced on a grassland river. Our destination was Rocky Ford Camp, where we'd left two trucks for the ferry back to our camp. The takeout was on river left just upstream from Rocky Ford Rapid, a class III hydraulic that our canoe—and whitewater skills—were not prepared to navigate.

On the last bend, the river constricted and swept us left toward the bank. We had to concentrate to avoid big rocks. The bank had been partially stabilized with fill rock to keep the river from swallowing a nearby road. We eased the canoe as close to the sandy loam of the curve as possible and could almost touch the shore at the outside of the arc.

Chloe made a whistle like the call of a black-capped chickadee (two descending notes, slight pause, followed by one note: chick-a-dee!) to get my attention. A mink climbed out of a hole in the bank and scurried up the duff onto a field of bluegrass. Mink are shy and reclusive by nature. This was a first sighting for Chloe, and rare for

me as well. Besides the Marais Des Cygnes River mink from a lifetime ago, I'd caught brief glimpses of them several times on the Kaw
and once while walking in the Wakarusa Wetlands back in Kansas.
We steadied the boat in the current and watched while the mink
scrambled through the grass to the edge of a mud slide it probably
used at night for sport. Pausing to sniff blades of grass at the top
of the slide, the mink was aware of us ten feet away in our yellow
boat. It made one more trip to its hole, then bustled over the bank
and into a plum thicket, disappearing from sight.

Our mink elation lasted for one nanosecond. Beyond the final
bend, the river made two tight meanders before straightening out
and barreling directly into Rocky Ford Rapid. After we cleared the
final curve, Chloe pointed left with her paddle. We needed to turn
the boat toward a two-story log building a few hundred yards ahead
of us. This required crossing the river.

The soft rushing sound of a rapid intensified, and a river-spanning
field of rocks blocked the channel ahead. Chloe swung portside
to aim at the takeout, but the current pushed us toward the riffle.
I tried a draw stroke to straighten the canoe, but overadjusted. We
slipped through a chute between the rocks and the white noise got
whiter. A final shelf of rock stood between us and smooth water.

We were perpendicular to the current now. When it was too late,
we both saw the boulder straight ahead. The canoe struck hard and
the current swung us around. The river began to swamp in through
two holes under Chloe's seat. I got out, steadied myself, and pushed
the wounded boat back into the stream, towing it the rest of the way
to shore. We borrowed a bailer and pailed out enough water to flip
the canoe and drain it.

I'm a careful planner and with a thousand-plus fresh grassland
river miles notched on my paddle, it felt discouraging that one brief
mental mistake could quickly end a voyage. At least we'd picked a
good place to bash holes in our new canoe: the tail end of a day on
the river, vehicles parked at the ready, the day before we were set
to return to Kansas.

While I bailed out the flooded boat, Chloe discovered another
clutch of *Gentiana puberulenta,* the downy gentian, also known

as prairie gentian. Even this late in summer, its deep blue-violet flowers were not yet blooming. This rare, though not endangered, beauty reminded me of another emblematic plant of Sandhills prairies south of the Niobrara. It was among the first things people who loved wild Nebraska told me about. Known as the rarest of all Great Plains forbs, it was thought to be extinct by 1940. For twenty-eight years, no evidence suggested its survival outside pressed specimens in herbariums until its rediscovery in 1968. Today, it persists in just thirty-two sites across the Nebraska Sandhills and Wyoming. A flower of isolated washes, too finicky for modern times, the black-footed ferret of grassland flowers, botanists have begun stashing specimens in seed banks in case all wild colonies vanish. Despite the dedicated work of biologists, it could still happen.

Rare plants inhabit rare places. True to its reputation, nobody could point me to a spot on a map to search for this botanical treasure. Chloe needed to get back to her classes in Kansas, but I felt compelled, even a little obsessed, to return soon, paddle more of the Niobrara, and seek out this mysterious plant. Perhaps, in searching for it, I would uncover a gateway into the timeless heart of the Nebraska Sandhills prairie.

Craig Freeman first encountered *Penstemon haydenii*, the "blow-out" penstemon, while working on his master's thesis in the 1980s. Now a plant systematist, botany professor, and senior curator at the University of Kansas, Freeman is a walking Wikipedia of grassland floristics. A key collaborator (along with thirty US and Canadian botanical institutions) to Oxford University Press's multidecade, thirty-volume *Flora of North America*[10]—the definitive taxonomic guide to plant life north of the Mexico border—his effusive positivity about all things native and herbaceous makes some people who meet him want to change careers and go into botany. Freeman penned the *Flora of North America* section on genus *Penstemon*, an endeavor that involved field work and the close examination of almost 25,000 specimens in herbariums across the United States.

Penstemon haydenii, sometimes called Hayden's penstemon, was named for Ferdinand Vandeveer Hayden, a geologist, explorer, and namesake of the nineteenth-century expeditions known today as the Hayden Surveys. He led an early government-funded scientific trip to document the Yellowstone Valley before it became a national park. In 1857, Hayden's team collected a specimen of an undescribed penstemon along the North Loup River and brought it back to the Gray Herbarium collection at Harvard University. This was the first cataloged example of *Penstemon haydenii*. Topping out at two feet, and taller than most Great Plains penstemons, the stems and blue-green leaves of blowout penstemon have a waxy armor to protect against blowing sand. Clusters of tubular lavender flowers appear from late May to early July and have a runway of dark yellow lines with golden hairs that guides insects to nectar inside the flowers. At one time, it was heralded as the only plant endemic to the state of Nebraska. When the Hayden Survey collected its specimen, blowout penstemon was probably common across the Sandhills, but that soon changed. The dunes of the Sandhills are "grass-stabilized" today, but, historically, mobs of grazing bison, periodic droughts, and fires routinely scrubbed the prairie veneer from large areas of dunes, exposing bare sand. The ceaseless winds of the open range were then free to bombard the naked dunes, excavating bowl-shaped depressions or "blowouts," often on the northwest sides of the sand formations. The process could take decades to produce stable "active" blowouts.

Penstemon haydenii is stickling about its habitat. It can only survive on bare sand. Like its early successional teammates, the blowout grasses (of genus *Redfieldia*) and lemon scurfpea (*Psoralidium lanceolatum*), blowout penstemon is a colonizer whose life cycle "heals" dune scars by establishing a rhizomal foundation other plants can build upon. Once other species become established, they quickly outcompete the shy forb and blowout penstemon dies out. Following the eradication of the bison, the cessation of cultural fires due to the forced removal of Native Americans from their ancestral lands, and improved range management techniques after the Dust Bowl, blowouts became rare as prairie stabilized nearly

FIGURE 7.2 Blowout penstemon (*Penstemon haydenii*). Photo courtesy of Craig Freeman.

every niche in the Sandhills. Freeman explained that "as vegetation became established and dunes became stabilized, the habitat that blowout penstemon occupied became scarcer and farther apart."[11] Habitat change was almost a death blow for the plant. By the 1940s, it was considered extinct.

In 1968, botanists rediscovered the species and set in motion a process that snatched it from the edge of oblivion. In September of that year, University of Nebraska biologists David Sutherland, Robert Kaul, and Dennis Brown found a population growing in a dune scar in Hooker County between Tryon and the Dismal River. At first, they didn't realize what they had found, although they noted that the basal leaves were unusually narrow and the flowered heads much larger than other penstemons. In a paper published about the rediscovery, Sutherland wrote that he found the plant "somewhat grotesque and commented that I wondered if it could be some sort of aberrant population, perhaps infected with a fungus."[12] The trio collected a single specimen. Back in his lab, Sutherland identified it from a 1930s field guide to Nebraska plants, published before the plant's precipitous decline. Sutherland didn't recognize the significance of his discovery until 1974, while reviewing a checklist of Great Plains plants, and noticing that *Penstemon haydenii* had been omitted because it was considered extinct. Meanwhile, a handful of other sites were found harboring the ghost penstemon.

Blowout penstemon had managed to hang on, and botanists declared it the rarest plant of the Great Plains. After wide-ranging searches, another population was discovered in a wall of sand dunes at the base of the Ferris Mountains in Wyoming. When Freeman and his colleagues performed molecular and morphological studies on the new population, they confirmed it was the same species, although discrete from the Nebraska populations and probably a separate variety.

The plant was added to the endangered species list in 1987, and eventually eleven native populations were discovered in the Sandhills. In the following years, populations have been bolstered with transplants germinated in greenhouses. As of 2024, there are likely fewer than one hundred thousand individual blowout penstemon plants growing at thirty-two known sites. At least one of those populations is at the 71,516-acre Valentine National Wildlife Refuge, located thirty miles south of the Niobrara River on US Highway 83. The refuge was established in 1935 to safeguard pristine Sandhills lakes, meadows, wetlands, and marshes that serve as breeding grounds for migratory birds and is home to 270 different avian species. It also conserves thousands of acres of prairie.

Mel Nenneman, biologist for the US Fish and Wildlife Service, works to preserve that prairie. Nenneman told me there are multiple types of grasslands at Valentine and in the Sandhills. "Near the lakes and wet spots where groundwater is close to the surface, we have what we call subirrigated meadows. That's where we have tallgrass prairie characterized by big and little bluestem and prairie sandreed."[13] Away from the wetlands, Nenneman said, the grasses become more mixed. "When you move into the Sandhills proper, which we have at the refuge, where it's more sandy soils, we get the mixed-grass or Sandhills prairie. On a lot of our uplands, we have both cool and warm season grasses. We have blue and hairy grama, wheat grass, those sorts of species, but also mixed in with little bluestem. It's definitely a hodge-podge that doesn't fit into a nice, neat little box." Like the rangelands of north-central Montana, 95 percent of the prairies at the reserve have never been plowed. Nenneman didn't say where to find blowout penstemon, but Valentine National Wildlife Refuge was known to harbor the plant.

The following May, I came back to Nebraska to paddle another seventy miles of the Niobrara. Afterward I drove down to the refuge. Photos from Google Earth revealed several large sandy yellow splotches in a remote section of the park. My car couldn't navigate the trail that led back into the hills, and I didn't have a tow rope. After getting as close as possible to the site, I stuffed the Google Earth printouts, binoculars, water, and a burrito into my red backpack and started walking down two tracks of sand heading toward open range, hoping they would lead to a blowout.

My phone was worthless in the Sandhills backcountry, the wind the only network with a dependable five bars. A sea of sand flowers bloomed between clumps of needle-and-thread grass and little bluestem. Yellow florets of puccoon, lavender blooming vetch, prickly poppy, sage the color of mint ice cream, wild rose with the last season's melted hips still clinging to thorny branches, small beds of prickly pear cactus, and yuccas ran to the tops of the dunes. I'd hoped to find painted milk-vetch, another unusual sand-loving plant, but it was too early in the season. The thin-stemmed forb doesn't set its red-mottled pods until later in summer.

The trail curved into a claustrophobic canyon of soft dunes, no two shaped alike, with steep slopes that hemmed the sky. Dune summits were covered in thin prairie like a middle-aged man losing his hair. A beetle the size of a watermelon seed rolled a piece of cow dung toward an anthill—Sisyphus recycling detritus. I stepped over the hollowed shell of an ornate box turtle, years dead perhaps, its best intentions starched white by the sun. Near a plum hedge, three sharp-tailed grouse flushed at close range and almost felled me from adrenaline. Eastern kingbirds, western kingbirds, and meadowlarks, speaking in unfamiliar tongues, flew between dead-wood yucca stalks. Layers of dunes that unfolded beyond the immediate canyon had the texture of clouds from years of polishing by the constant prairie wind.

Two miles in, I reached a windmill in a protected cove of bluffs and, below it, the lower quadrant of a dune sheared off to expose a cavity of pure sand ten feet below the curve of the hill. The same yuccas and puccoons and roses ringed the green prairie around the blowout, but fewer plants grew in the scar. I carefully walked around the edge and noticed a flower with vivid cerulean blooms near the bottom of the wash. Surely, it must be blowout penstemon. The basal leaves were short and alternating, with a waxy hairlike membrane on the stems and leaves. The stalks of flowers and petals were almost psychedelic in hue, blue and lavender, like they had been hand-painted with horsehair brushes dipped into jars of compounded robin eggshells and blush wine. Despite their beauty, the flowers had little fragrance.

Ecstatic, I knocked on a yucca stalk in honor of my dumb luck. I couldn't believe it had been so easy to find America's rarest prairie flower. Careful not to disturb the sands of the blowout, I spent an hour taking photos, luxuriating in the rare habitat, and marveling at the moody blue blooms that almost didn't seem real.

I kept on the lookout for more penstemon on the hike back. It didn't take long to find some. Penstemon was blooming in the pure sand ruts of the trail. On the drive out from the refuge, a squadron of pelicans anchored in the water of Pony Lake, beneath a dead tree where four cormorants slept in the branches—ivory and ebony. Getting out of the car to snap a photo, I saw more penstemon plants waving the bandannas of their blue petals.

Crossing the bridge over the Niobrara later that evening, a green blanket of prairie beneath the "Welcome to Valentine" sign blazed with sand flowers. I pulled over to take a closer look and found a healthy sprinkling of penstemon. This was where the city rolled out the red carpet to visitors. Perhaps this landscaped prairie was meant to show the world that Valentine was the gateway not only to the Niobrara, Nebraska's rarest and most iconic grassland river, but also blowout penstemon, Nebraska's rarest and most iconic grassland forb.

It was not meant to show that.

Craig Freeman wrote accounts of 240 North American penstemons, and the one I found at Valentine National Wildlife Refuge and the bridge over the Niobrara was narrow leaf, not blowout, penstemon. The vibrant green prairie on the city limits of Valentine was a *prairie*, not a blowout. Narrow leaf penstemon is a wonderful native flower of untainted prairies, yes. It is a proud line item in Freeman's account of the penstemons of North America, for sure. But *Penstemon angustifolius* is a commoner when compared to the critically endangered blowout penstemon—a prairie dog, not a black-footed ferret.

Knowing there were at least thirty-one more known sites to look for the plant, I figured my next best chance would be on public land about an hour west of Valentine, where I'd heard of a purported population. Back in my motel room, Google Earth showed a large yellow blotch that resembled a child's sandbox in a remote part of the preserve. It was promising, yet perplexing. The dunes were in the middle of one of the largest national *forests* in the Great Plains, not where you'd expect to find virgin grasslands. But Nebraska, and Nebraskans, have a long and strange history with trees.

Nebraska was once quintessential grassland country, with grasslands covering 98 percent of the future state. When the first European Americans began to settle the prairies and Great Plains, they planted nonnative pines and junipers around their homes for comfort and protection from the wind. However, these shelterbelts were mere accessories compared to Julius Sterling Morton's ambitious vision for trees in future cities of the Midwest. In 1872, Morton, who was born in New York and grew up in Detroit, founded Arbor Day in Nebraska City (the wealthy newspaperman and future US secretary of agriculture also fought to keep slavery legal when he served as secretary of the Nebraska Territory prior to the Civil

War). The Arbor Day Foundation sponsors the "Tree City USA" program to promote the benefits of trees in urban environments. In 2024, ninety-one Nebraska cities were on the Arbor Day Foundation's list of tree cities. In addition to Arbor Day, Nebraska is home to two national forests . . . of *planted* trees.

Morton might be better remembered, but the real champion of Nebraska forestry was Charles E. Bessey. Born in a log cabin, the Ohio native was professor of botany and horticulture at the University of Nebraska from 1884 to 1915. Bessey believed that state land-grant colleges should serve the public good, "science with practice" as he liked to say. Toward that end, Bessey established an agricultural field experiment station, sorted out problems in the field of modern plant classification, served as president of the American Association of the Advancement of Science, and promoted sustainability and conservation practices. He authored the textbook *Botany for High Schools and Colleges* and founded one of the oldest and largest herbariums in America, named after him today.[14]

Shortly after arriving at the University of Nebraska, Bessey and his students immersed themselves in botanical studies of the Sandhills. Bessey was interested in land use practices that could benefit the state's agriculture industry. Ranchers in the treeless Sandhills needed wood to build cattle shelters and fences. Demand for timber had risen nationwide as the country expanded westward. In 1887, Bessey helped campaign for a state tree preserve in the Sandhills, a place with almost no trees. Although the Nebraska legislature balked at the notion, Bessey thought that the ample subsurface moisture below the dunes would support healthy pine forests. In 1891, the US Division of Forestry oversaw the planting of 13,000 conifer seedlings—mainly ponderosa pine, jack pine, red pine, Scotch pine, Austrian pine, and Douglas fir—on a small ranch beside the Middle Loup River near Chadron.

After planting those first trees, Bessey's experiment languished for ten years and was almost forgotten. When the Division of Forestry asked for a project update in 1901, Bessey and his students surveyed the ranch and found tracts of eighteen-foot-tall ponderosa

pines. This proof-of-concept convinced US Forestry director Gifford
Pinchot to urge President Teddy Roosevelt to act. Nebraska National
Forest was created by presidential proclamation in 1907 and today
occupies a combined 200,000 acres in the remote Nebraska Sandhills. A separate preserve, the 116,000-acre Samuel R. McKelvie National Forest, is located on the south side of the Niobrara River in
Cherry County. Despite its name, the Samuel R. McKelvie "forest"
is mostly a wild Sandhills grassland, one of the finest large sweeps
of public prairie in America.

I had to hustle. It was late in the day when I arrived at Samuel
McKelvie and in two hours the sun would dip beneath the tawny
hide of prairie-covered dunes. The wind was up, pushing against the
driver's door. Parked in a circle of ATV ruts in raw prairie, I stuffed
an unfoldable national forest paper map and binoculars into my
backpack, along with provisions from my motel room. Without
camping gear or a headlight, I'd rely on mental breadcrumbs to
find my way back from the blowout—wherever it was—if night fell
and I was out there alone in the grasses.

Hopscotching the cattle guard, I began following a trail that soon
led into a striking prairie, one of the finest I had ever seen. Glanc-
ing back to make sure the headlights blinked when I pressed the
remote button, an indigo-speckled field of narrow leaf penstemon
surrounded the car.

That the blowout was not on the main trail was certain. Where
to leave it and strike out for the great sandy mark on Google Earth
was less so. I figured the blowout was two to six miles from the car,
depending on how serpentine the path became. I knew I should
walk fast, but the prairie was entrancing, a siren. I whistled back at
meadowlarks and stopped to study the fractaline patterns of micro-
cacti bricked into thorned gardens. The humming of life among the
sand flowers was dizzying—the yuccas and the needle-and-thread
grass; the little bluestem and puccoon flowers; the auditory ping-
pong of bird calls zinging by in all directions; the wild raspberries
and strawberries; the lumbering box turtles; the millennial continuum of wind pulsing through the biomass; the hills as sensual as

the curve of a hip, prairie dog poetic; the grasses' strumming current testament to why it was once described as *sea*.

Two miles in, coyote tracks led to a badger den at the top of a ridge. Glassing with my binoculars from the high point revealed no signs of the blowout, but the blades of a windmill spun on a little rise about a mile down trail. A half mile beyond the windmill, a faint footpath split off from the main trace and led south toward a line of dunes. It could lead to the blowout, so I followed. A nighthawk buzzed low, a reminder of impending darkness. It flew toward a lone tree but then swung sharply back around. A red fox walked across the round buffalo wallow beside the trail, ignoring me as it trotted off toward the peeled dune ahead of us. There was nothing to do but follow.

When I finally reached the sandy dune, my spirits sank. The blowout was desiccated and gloomy compared to the prairie I'd crossed. Two acres of bare sand filled the sheared-off face of the dune.

But blowout penstemon was everywhere. There was no mistaking it for its narrow leaved cousin. The waxy plant dominated the dune scar, like cocklebur on a sandbar. The plant only grew in the heart of the sand. It was taller and bulkier than narrowleaf penstemon, and even this late in May, not blooming yet.

Penstemon haydenii is a colonizer. It kudzus across bombed-out dunes, gathering forces as if to cry out among the meadow-larks: *Prairie is coming!* The penstemon looked tough enough to endure anything that the Sandhills climate could dish out. Hundreds of dunes were visible from the top of the wash. In the cosmic arc of Sandhills evolution, I wondered if each dune's prairie once sprang from a single wind-delivered penstemon seed in thousands of individual grassland big bangs.

Careful not to walk too near the plants, I sat in the grass at the top of the blowout. The penstemon were arranged in seven colonies, each with about a dozen individuals. It was humbling to sit alone in the wildness of the place, one of a handful of spots that hopefully were not the "last stands" of the rare plant. The Sandhills were not

always prairie stabilized. For thousands of years, the boundary between bare sand and prairie ebbed and flowed like the tides. Oregon trail travelers reported walking from Valentine to North Platte
without stepping on a single blade of grass.

That's what Greg Wright told me. Wright is a US Forest Service
wildlife biologist who specializes in endangered species, a field he
was drawn to so he might "impact life on earth" as he put it. Wright
has worked on whooping cranes in Nebraska and Texas and chi-
nook salmon in the Pacific Northwest. Since moving back to his
native Nebraska, one of Wright's charges is blowout penstemon at
Samuel R. McKelvie National Forest. When we spoke two weeks after I found the blowout, he told me penstemon plants there were
reintroduced, but the plant was native to the vicinity in historical
times. "In the 1800s, this was a quite different habitat, where blowouts moved across the landscape like sandbars on a river, one in-
tersecting another. We can assume that anywhere in the Sandhills,
the species could have existed. Collection back when the plant was
common wasn't prolific, and there are no known records of blowout
penstemon from McKelvie. But the habitat is still here."[15]

Beyond anecdotal evidence, physical records support his claim.
"I've looked at aerial photographs from the 1950s from the Bessey
ranger district down at Halsey on the Middle Loup, and the amount
of bare sand from those first aerial black-and-white pictures are
like night and day compared to now. You can only imagine what it
was like in the 1800s when fire was routine across the landscape,
there were no trees whatsoever, and roving herds of bison came
through that took three days to pass. There were a lot of different
influences on the landscape then. Fire and the grazers were the two
biggest ones."

Each blowout penstemon plant produces about 1,500 hard brown
seeds that look like tiny seahorses. Only a third of the seeds are
viable. They can survive for decades beneath the sand. Wright said
that after the plant's rediscovery, scientists had to devise a recovery plan based on just a few thousand individuals. Botanists at the
University of Nebraska in Lincoln harvested seed and conducted ex-

periments to figure out the best way to establish new populations. At first, they propagated plants in greenhouses. Over time, they established new sites and the seed base expanded.

In the wild, sand blasting from wind scours seeds and removes their protective coatings. Penstemon seeds need about two weeks of steady moisture to germinate, with ideal conditions occurring only once every eight to ten years. Since the plant lives only six years, existing populations are at peril and require active conservation.

Interventions by biologists like Wright are helping at this crucial stage in the plant's recovery. Now that more seed is available, Wright has experimented with direct seeding rather than relying completely on greenhouse transplants, a strategy that leverages the plant's symbiotic relationship with blowing sand. "A blowout is essentially a big wash tub, or more accurately, a dryer, where seed gets mixed around and sandblasted. The big conical shape of the formation traps seed. Then nature chips away at the coating and prevailing winds and hoof action and all this other stuff moves the seed into the places where it's most likely to succeed. By doing that, we're actually reducing risk, because the plant is not starting out where *I* think it needs to go. It's starting out where the natural system places it."

This approach helps Wright and others learn what habitat the plant prefers, further eliminating guess work. "If I sprinkle a seed out there and it doesn't turn into a plant, then something's wrong, something's off. It's telling me as a biologist that it's not in the right habitat. I've started pairwise experiments with different shaped blowouts, different sized blowouts, different amounts of seed, and different seed planting treatments to try to figure out what the nuances are."

A cattle producer might consider a half acre of bare sand unproductive acreage, but for blowout penstemon, scientists are finding that a half acre is the bare minimum. Bettering the odds of the species' long-term survival might require finding larger blowouts of several or dozens of acres. Wright said, "Stochastic events like years of really great rain can snuff out a half-acre blowout a lot easier than a six-acre blowout. We're learning that our perception of what a big blowout is might have been wrong at first. Bigger blowouts seem to be more appropriate for the species to be able to really do well."

The difference between a half acre and six acres is negligible compared to the habitat requirements of many endangered species. This bodes well for the future of blowout penstemon. Wright said, "We're not trying to rewild and convert twelve million acres to bare sand, to re-create an entire biome. We're looking for five acres here, twelve acres there. Eastern red cedar is a much bigger ecological threat to the Sandhills than bare sand."

Write told me he believes there is a philosophical aspect as well: "I think we're starting to realize that sometimes things are just important for the sake of their own existence even though we might not understand their importance as a species right now. Maybe in the future we will. Or maybe it's just cool that something that's been around for thousands of years can continue to exist. Blowout penstemon is a beautiful plant that belongs to this landscape, even if only in moderation. It's the kick-starter that gets prairie going on a bare dune, so we know it has an early successional role. But it's just nice for us to have a few big blowouts with penstemon where we can go and take the kids and show them the rare flower."

Since the Sandhills formed thousands of years ago, the environment has kept them moving along a continuum between bare sand, like the Sahara Desert, and prairie-stabilized sand that dominates today. A wildcard that could shift the balance toward a sandier future is climate change. The 1930s proved that long-term drought encourages new blowouts to form and existing ones to expand. Wright said, "This is a plant we might need one day, if you want a practical reason to save it."

Saturday night, I was alone on a remote sand dune, beside a penstemon patch rarer than ball lightning, in the middle of deep geologic time. Emily Dickinson wrote, "To make a prairie it takes a clover and one bee." If she had lived in the Nebraska Sandhills, she might have written, it takes only *Penstemon haydenii* and time.

From Rocky Ford Rapid, where we drowned the canoe, the Niobrara flows southeast across most of Nebraska before veering north again to meet the Missouri River near its final impoundment at Lewis and Clark Lake. From there to St. Louis, the Missouri is undammed but not ungirdled. The big river leaves the lake and flows east through Yankton, South Dakota, then southeast to Sioux City, Iowa, on its way south. Below Omaha, it's almost unrecognizable as the river of the Upper Missouri Breaks National Monument, its younger self.

The lower Missouri gathers the water of thousands of grassland rivers, bringing them home to the Mississippi and eventually the Gulf of Mexico, where they either sink into the nameless depths or evaporate into clouds that hitch back north toward the plains to begin the process anew. Channelized like no other grassland river, the lower Missouri is throttled with wing dikes that guarantee barges a consistent navigation channel. The 340-mile stretch from Kansas

FIGURE 7.3 The Middle Loup, a classic Sandhills river in central Nebraska. Photo courtesy of Lisa Grossman.

FIGURE 7.4 The author and his daughter Chloe (*left*) with the Kaw River guides, Niobrara Wild and Scenic River. Photo courtesy of Dawn Buehler.

City to St. Louis embodies the paradoxes and possibilities of America's grassland rivers. I was eager to experience it in one epic three-day binge—along with five hundred fellow paddlers who had signed up for the Missouri River 340 ultramarathon race, the world's longest nonstop river race and yet another thread in the modern story of grassland rivers. First, however, I had one last stop to make along the Niobrara.

During his 1861 journey to Minnesota to seek relief from the tuberculosis that soon killed him, Henry David Thoreau sauntered through a tallgrass prairie, marveling at a landscape completely devoid of trees. The chaste transcendentalist from Concord, Massachusetts, was one of America's first nature philosophers. Thoreau believed that humans had a spiritual need for the "tonic of wilderness."[16] A hundred years later, a side note in the Wilderness Act of 1964 provided a legal definition of Thoreau's tonic. One of the act's principal authors, Howard Zahniser of the Wilderness Society, defined wilderness as a place where "man himself is a visitor and does not remain."[17] Since then, more than 109 million acres of federal land have been protected by the act. Congressionally designated wildernesses disallow not only those who *would remain*, but also their motorized vehicles, mechanical equipment, and roads. Wildernesses enjoy the strictest protections of our public lands.

Few of the more than 800 federal wildernesses include low-elevation midland prairies. Hercules Glade Wilderness in Missouri has a small tallgrass prairie. The Badlands Wilderness in Badlands National Park includes thousands of acres of shortgrass prairie where black-footed ferrets have been reintroduced. Fort Niobrara Wilderness, in the Fort Niobrara National Wildlife Refuge, is the only federal wilderness that enshrines a grassland river.

I asked refuge manager Matt Sprenger if there was anything unique about the 4,000-acre Fort Niobrara Wilderness that set it apart from the rest of the Sandhills. I was interested to see how he thought the site fit into the American wilderness gestalt. He said the designation was more historical than philosophical. "The land in the wilderness went from a military installation directly to the Fish and Wildlife Service. Nothing was ever developed on that land. It was fairly pristine to begin with, like a lot of the other canyons. Since it was important, along with the rest of the refuge, for the conservation of native birds, it remained undeveloped. When they were looking for areas to designate wilderness back in the 1970s, this was the strongest candidate in Nebraska already being managed at the federal level."[18]

The trail to the wilderness starts at pens used to corral 350 bison who spend much of their time on the wilderness. Bison have been managed at the refuge since 1913, so they were only missing from the Niobrara landscape for a few decades. The Fort Niobrara Military Post operated between 1880 and 1906, a part of US military strategy of the time to "surround and contain" the Lakota people after the Black Hills War of 1876, a series of skirmishes and negotiations between the United States and bands of the Lakota Sioux and Northern Cheyenne. Seven years after the fort was abandoned, 16,000 acres became Fort Niobrara National Wildlife Refuge, soon protecting not only migrating birds but bison and elk as well.

The hike up to the south rim above the river was steep. When

I closed the bison gate behind me, the metal latch rang like a church bell. And, in that moment, it was a church. I placed my backpack on the ground and stood in the middle of the prairie, trying to absorb the theology of the landscape: to the east, a solid wall of short oaks beside the pasture where the canyon fell into the valley carved by Fort Falls; to the south, lavender prairies ripening on soft dunes; to the west, the pastures of the preserve and the wide Niobrara itself, engorged by Cornell Dam; to the north, ponderosa pines standing their ground, unmoving in the eternal winds blowing off the Sand-hills, the crowns of Ice Age aspens and birches piercing the over-story.

Oh, Nebraska! The Dismal, the Middle, South, and North Loup, the Missouri, the Platte, and, here, the Niobrara, the river that brought me to a true prairie wilderness. Anemic by wilderness standards—the Eagles Nest Wilderness close to urbane ski communities in Summit County, Colorado, checks in at 133,000 acres—at least one pasture in the prairie world I so loved was not only wild, but wilderness. A tonic to my spirit at least.

Although the designation might cause refuge managers head-aches (fire crews have to manage controlled prairie burns without the aid of mechanical equipment), the classification serves a higher purpose. Whether the refuge ever needs a new visitor center, or a ranch changes ownership, or the control of an elected office switches political parties, this never-plowed prairie and the six ecosystem elements along the river are off-limits to development. Maybe that's enough. Most prairies in middle America, after all, have been long since plowed under.

Acknowledging that, per congressional decree, this prairie was not a place to remain, I slung the backpack over my shoulder and started hiking down toward the buffalo pens, breathing deep the timeless smell of summer and crushed prairie sage.

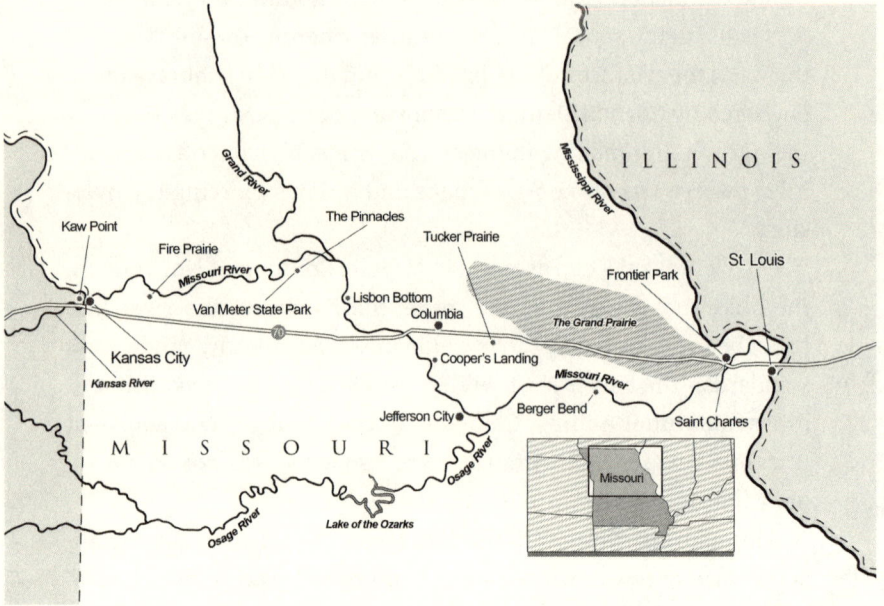

MAP 8 The Lower Missouri River

CHAPTER 8

MISERY

Waiting for the sun. A solemn procession of paddlers and crews lugging half-loaded kayaks, canoes, stand-up paddleboards, and pedal-drive boats formed a silent queue from the riverside forest down to the boat ramp at Kaw Point. When it was our turn, we waded the *Getting to Nowhere* into the carob-hued waters of the Kansas River. Christina helped lash three dry bags into their appointed positions, jimmied the water jug into its foam holder behind the seat, velcroed the drinking tube onto my lifejacket, and pushed me and the canoe off into the urban slurry.

After paddling upstream to the Lewis and Clark Viaduct—where explosives had dropped the old bridge on my last visit—I half-mooned the canoe and took my place among 419 other boats awaiting the start of the MR340 ultramarathon kayak and canoe race. Two football fields ahead of our flotilla, the Missouri River was swallowing the facile waters of the Kaw like a tornado inhaling a trailer park.

The lower Missouri is a legendary river. Inhabited for at least 12,000 years, prior to Euro-American settlement, it was home to the Missouria, the Kaw, the Osage, and other Native American groups who spoke Dhegihan and Chiwere dialects of the Siouan language. The French were the first Europeans to claim control of the valley they called "Upper Louisiana." St. Louis, named after the thirteenth-century canonized king Louis IX of France, was founded near the mouth of the river in 1764. A period of Spanish rule followed before France regained nominal control again for two years in 1801. Daniel Boone, the famous but broke frontiersman who had grown disillusioned with the fledgling United States, expatriated himself to Spanish-controlled territory near the big bend of the Missouri in 1799. Following the Louisiana Purchase of 1803—which included the entire Missouri River valley—the United States claimed possession and American scientific and military expeditions began traveling upriver, documenting its cultures and environs.

Before railroads, the Missouri River was a primary transportation route of expansion west of the Mississippi. From 1819 until the end of the nineteenth century, wooden side-wheel steamboats navigated the Missouri's precarious shoals, sometimes backing their sterns onto sandbars or cornfields to make tight turns at quixotic bends in the river. Most steamships in early years belonged to the American Fur Company, based in St. Louis. By the 1850s, luxurious four-floor "floating palaces" with saloons, private staterooms, and chamber orchestras plied the river, traveling upstream at ten miles per hour, carrying hundreds of tons of cargo and ticketed passengers. By the 1860s, competition from railroads and a fledgling barge industry cut into freight revenue, and the steamboat era began to wane. Before 1900, an estimated 300 to 400 ships sank in the river, many still buried beneath cornfields in forgotten channels carved during floods.

Radical environmental change followed the Civil War. To create a navigable waterway for barges, the US Army Corps of Engineers rewired the slow-braided Missouri, creating a deep fluvial highway locked into a fixed channel. Once sinuous and free to stalk its enormous floodplain—ten miles across in places—the lower Missouri had previously flowed through a vast kingdom of tallgrass prairie. Those eastern prairies, with few exceptions, have vanished, converted to croplands in the fertile bottoms surrounding the river.

Growing up in the suburbs outside of Kansas City, I'd dreamed of paddling across the entire state of Missouri; but I worried the big river would be too much to handle alone. The MR340 was my best hope to take on the lower Missouri in one shot. The race begins at Kaw Point—the historic Lewis and Clark encampment in the shadows of downtown Kansas City—and ends at the Lewis and Clark Boat House in St. Charles, where the expedition began their journey of "discovery" through a land that had already been mapped by the French and Spanish and known for thousands of years by indigenous peoples.

I'd trained hard for the race, paddling the 340-mile distance twice in biweekly training runs leading up to the mid-July start. I didn't care about times or *racing*, but the grueling event required careful preparation to complete safely. The MR340 has a strict eighty-six-hour limit for solo boats and interstitial checkpoints with their own cutoff times along the way. Paddlers must beat a boat—*The Reaper*, with its skull-and-crossbones flag—to each checkpoint or be disqualified. Nowhere in my quest to learn about grassland rivers had I faced a cutoff time, other than sunset. For that, I brought a tent.

Each race team must also have a support crew. Because none of my friends were eager to take a week off work to slog across Missouri catering to my blistered, spaced-out paddling needs, Christina had volunteered for the job. Neither of us really understood what that entailed. We found out later that "ground crew" usually meant a husband, wife, brother, sister, or parent. I had made a rough map of potential stops during the race and general notes for what we'd need to restock along the way, but the map was vague and unworthy of the word *"plan."* Christina had never driven across Missouri by river road—or *anywhere*—towing the small teardrop camper we were bringing.

The sun began to rise behind the buildings of downtown Kansas City. The festival of paddlers, ground crews, and their vehicles would soon begin moving in one large wave across the state of Missouri. A DJ, who had been blasting classic rock and disco, spun a scratchy rendition of the national anthem. Afterward, I raised my paddle overhead and made a quick prayer that I could beat *The Reaper* to the first checkpoint, seventy-four miles downstream in Waverly by dinnertime.

In addition to provisions to help me complete the race, I brought along years of questions about the environmental conditions of the big river and its prairies. The Missouri had suffered a slow century and a half of hydrological insult. How would this landscape compare to the immense intact prairies of the upper Missouri in north-central Montana, or the Niobrara in the Nebraska Sandhills, or the sweeping miles of buffalo grass along the Purgatoire and the South Platte, or the hard-fought tall prairies of the Kansas and Grand Rivers?

Over the next four days, I would discover a scarred river, yes, but a river cherished by a devout ensemble of protectors: paddlers and boaters, birders, anglers, hunters, barge operators, cyclists, farmers, historians, musicians, and many more. I would also find that the lower Missouri was haunted by prairies past, present, and future.

Prairies past: like the Fire Prairie, the Fox prairie, the French prairies of St. Louis, the Petite Saw Plains, and the Pinnacles, a complex of loess bluffs once covered by a vibrant tallgrass prairie. Partially contained in Van Meter State Park, for hundreds of years the Pinnacles were home to the Missouria people and their ancestors. The park protects a centuries-old earthwork of ditches and embankments at the site of a Missouria village, a vivid reminder of what life along the lower Missouri was like for hundreds of years.

Prairies present: like the last fingerling remnants of the Grand Prairie, which once stretched west from St. Charles, and the floodplain prairies within Big Muddy National Fish and Wildlife Refuge, a collection of protected bottomlands that the US Fish and Wildlife Service is stringing together one unit at a time to form an archipelago of reserves from Kansas City to St. Louis. Created after the near-biblical floods of 1993 to protect and restore wetlands, scour holes, riparian forests, and grasslands, the refuge could one day give the river room to find its natural rhythms again.

And prairies future: because, as the Missouri nears the Mississippi, native grasslands vanish. Because these eastern prairies disappeared early, the prairie region along the Mississippi River is the cradle of grassland restoration science. The oldest restored prairies are nearing their centennials. They have yielded a botanical trove of data that, along with recent discoveries and breakthroughs, is helping wildland managers understand the most effective ways to reconstruct the mind-boggling biological diversity of indigenous grasslands.

For now, though, fluvial reality was all that mattered. This moment had looped hi-def in my mind for weeks. The throng of paddlers lurched forward in one great heave toward the wall of moving water that had started as melting snowpack in northwestern Montana before traveling two thousand miles across the grasslands of America. Aiming the bow at the tip of Kaw Point, I dug in hard to generate an escape velocity sufficient to avoid being flipped by the great current of the Missouri when it struck the Kaw. The three kayaks barreling forward ahead of me caromed to the right and disappeared like they had been struck by a billiard ball. At the cusp of the confluence, riding the Kaw's final breath of current, I held on and felt the canoe, and my fears, bash into a solid wall of river.

The Missouri River Commission was created by Congress in 1884 to accomplish "continuous, progressive development of the river" for navigation and flood control purposes.[1] This was the beginning of the end for the old river. At the time, the Missouri was free flowing and undammed from its source in the Montana Rockies to its mouth at the Mississippi. During the steamboat era, the Missouri River's vicissitudes drove up the cost of transporting passengers and raw goods. In 1867, army engineer Charles Wagoner Howell counted 1,793 snags on a trip upstream from St. Louis by steamboat.[2] The goals of the engineering-oriented government agency included creating surveys, stabilizing banks, clearing snags, and establishing a minimum depth sufficient for commerce. In other words, taming the river.

Between 1878 and 1885, the US Army Corps of Engineers surveyed the river and created topographical maps that were used by work crews to place heavy timbers and create wooden pile dikes that diverted the current away from the banks and into the middle of the river. By the first decade of the twentieth century, an intricate lattice of fluvial strictures was in place from Rulo, Nebraska, to St. Louis. When the work was complete, laborers had removed 17,676 snags, 6,073 fallen trees, and sixty-nine drift piles from the previously wild Missouri.

Meanwhile, the Corps created plans for a series of reservoirs for hydropower, municipal water supplies, and irrigation. Completed in 1891, Black Eagle Dam near Great Falls was first. Over the next sixty years, the Corps built six mammoth lakes on the main stem of the Missouri, further altering the nature of the river downstream.

In the twentieth century, two congressional acts were approved to further convert river to highway. In 1912, Congress authorized the Corps to construct a six-foot navigation channel on the lower Missouri. The project was finished in 1933 at a cost of sixty-eight million dollars, even though the volume of goods had dwindled to a third of the tonnage transported at the peak of the steamboat era. The Rivers and Harbors Act of 1945 authorized the Corps to deepen the river yet again, this time to provide a 9-foot deep, 735-mile navigation channel below Sioux City. The project took thirty-five years to complete. Sand and gravel mined from the river was used to construct wing dams to increase stream velocity and allow the river to self-scour its channel. Banks were lined with rock to reduce erosion and "fix" meanders that used to rearrange themselves in high water. Levee systems were built to shield farms from floods. The sum of these changes made the river narrower, deeper, and faster than in pre-European settlement times. Once a symbol of untamed wilderness, it had become a shadow of its former self, unrecognizable to the dwindling few who remembered its original, free-flowing nature.

After opening my eyes to find the trip through the confluence hadn't flipped me, the Missouri slung the canoe hard right, and I joined the conga line of bobbing paddlers hugging the right bank below the Kansas City flood walls. The line bulged left around Woodswether Terminal, the only working barge port in Kansas City.

Calming into a methodical groove, I began passing crafts that were fading after a fast start: a man in a twelve-foot blow-up kayak he told me he'd bought for his daughter at Walmart, a woman in a banana-yellow kayak named *Pattitude* who was biting off lengths of blister tape, and three stand-up paddleboards, boats that the canoe and kayak people didn't quite understand. "SUP" paddlers lash their gear to the board and employ long-handled paddles in a "stand and switch" propulsion technique. When the wind comes up, they can kneel on the board or sit, often on milk crates. One woman was blasting the song "Daft Punk Is Playing in My House" by LCD Soundsystem from dayglo orange polyester "fur"-covered speakers attached to her crate. But the great diversity of kayaks was unrivaled among the other boats. There were thin speedsters that looked like dental implants propped up with outboard riggers ("training wheels" one guy called them), heavy plastic subs jury-rigged with fiberglass patches, couples with matching his-and-hers kayaks, sit-on-tops, sit-insides, whitewater kayaks designed for tight turns that seemed ill-suited for the Missouri, kayaks with extra stabilization for nature photography, sea kayaks, one kayak with a transparent plastic bottom for viewing coral reefs, kayaks with huge storage capacities in sealed bulkheads, and one with no storage, its racer carrying her gear in a big fanny pack. To spectators lining the walkways, we must have looked like a migration of geese; groups streaming out into V formations, trailing boats drafting the bow wakes of the lead boats as they broke the current. To beat *The Reaper*, I couldn't afford even one stop to get out of the boat and stretch my legs. As race founder Scott Mansker urged in one of the prerace emails, I tried to visualize a rope attached to my life vest, pulling me toward Waverly.

MR340 founder Scott Mansker learned to paddle big rivers in the 1980s. One time, he and a friend launched a raft they had built from Styrofoam logs and plywood onto the Missouri in Yankton, South Dakota, for the 400-mile journey to Parkville, Missouri. In the first five minutes of their trip, they crashed into a dock, destroying the raft. Instead of giving up, they sold it for parts and continued in a square-back Grumman canoe.

After getting to know the river firsthand, Mansker felt that the lower Missouri was unlike any river in America. It had a town with a boat ramp every fifteen to twenty miles, many with restaurants, grocery stores, and riverside parks that could serve as staging areas where support crews could meet paddlers. The lower Missouri would be the ultimate river for an ultramarathon paddling race. He'd heard of the Texas Water Safari and the Yukon River Quest, long races that attracted thousands of paddlers. But the Missouri River between Kansas City and St. Charles, the historic river town twenty-nine miles from the Mississippi River on the edge of St. Louis, had one leg up on these other races: the Missouri has no river-spanning impediments that would require a portage. There were no lakes, dams, or diversions on the course Mansker had been grooming in his mind—just miles of wide-open grass-fed river.

In 2006, he launched the inaugural Missouri American Water 340-mile river race by putting a web page up and telling some friends about it. The first year, when the race drew about twenty entrants, one of his stated goals was that nobody should die. Since then, the event has grown to almost a thousand participants. In 2020, the grassroots group Missouri River Relief, whose mission is to connect people to the river, took over operations. As of 2024, Mansker remains intimately involved, an ultramarathon river shaman to paddlers drawn to the Iditarod of American ultramarathon paddling races. Organizations like Midwest Paddle Racing put on dozens of shorter races as warm-ups for the main event.

Four hours after leaving Kaw Point, the navigation channel, which for miles had followed cornfields in the Liberty Bend Bottoms, crossed to river right. I paddled hard across ten-foot boils—smooth bubbles that form on the surface, undulate for a while, and then pop, only to reform again. Black waves with frothy tops piled up against my port side, rocking the boat. Cottonwood snow fell and gathered in drifts closer to shore. The sun rose higher in the sky and, for the first time, it started to feel hot. I took a watermelon salt tablet from my life jacket and chewed it while keeping the boat moving forward.

I found myself trying to stay within audible range of boats rigged for music—mainly free-spirited SUP dudes playing songs that guys who spend a lot of time with their dogs listen to. I'd never listened to music on a river before, not even through headphones. But if these riverine DJs had stapled tip jars to their boards, I would have chipped in. The diversion helped me keep up the pace.

Not far beyond the mouth of the Little Blue River, we reached the unofficial paddle stop at Fort Osage, the trading post and military fort established in 1808 that was rebuilt from preserved surveys in the 1940s. A few paddlers crossed the channel to meet with ground crews, but I kept plowing forward. The river widened and slowed on the way toward the old village of Napoleon, a town five miles away from Wellington. Named for the commanders of the final battle of the Napoleonic Wars, the two villages were founded within a year of one another in 1836 and 1837. The community of Waterloo was laid out between them in 1905.

The river veered right beside a plain of crops and scrubby forest that was once a well-known grassland called the Fire Prairie. In an 1812 trip with the Missouri Fur Company led by Manuel Lisa, John C. Luttig wrote in his journal that the expedition killed seven deer, one beaver, one fox, one pelican, one turkey, and five bears— two adults and three cubs—while camped in an open grassland at the mouth of Fire Prairie Creek.[3] At the same spot seven years earlier, William Clark and John Ordway journaled entries that could have been mistaken for real estate advertisements. The "River of the Fire Prairie," Ordway wrote, "has a butifull [sic] high bottom for 1½ miles back and rises to the common level of the Country about 70 or 80 feet and extends back out of view."[4] The expedition killed a five-hundred-pound black bear along the creek, one of a dozen bears they shot on the prairies within fifty miles of Napoleon. The grasslands of the Missouri River along the Fire Prairie once teemed with bears.

I counted four deer, four turkeys, two beavers, and one gray fox between Fort Osage and Napoleon, but no pelicans, no bears, and, when I paddled by the mouth of Fire Prairie Creek, no fire or prairie, although the stream runs close to a fireworks store. (Most Missouri creeks probably flow within a mile of a fireworks store. Per capita, the state has led the nation in fireworks spending for decades.) I'd seen white pelicans in Upper Missouri River Breaks National Monument, and they still frequent the lower Missouri. Nobody, to my knowledge, has seen a black bear during the MR340. That may change. In 2021, the bear population of Missouri had grown to one thousand, and the Missouri Department of Conservation (MDC) established the first modern black bear hunting season in the history of the state. Most black bears live in the southern counties of Missouri but, every year since 2010, there has been at least one report of a black bear north of the Missouri River.

The last miles to the checkpoint were a glycogen-depleted blur. The breeze making water music in the cottonwoods, the purr of the river, and the aches in my shoulders pushed me harder toward temporary relief. Taking no chances when the bridge came into view, I sprint-paddled to the finish, clocking in at the checkpoint at 6:29 p.m., one hour and thirty-one minutes ahead of *The Reaper*. At the boat ramp, two volunteers wearing swim trunks waded waist-deep into the river and pulled my boat up onto the sand. My legs flowed like liquid when they helped me up. "Lean forward," one said. The ramps have a ten-degree pitch and it's better to fall forward, if you must, than to crack your head on the concrete.

The park at Waverly was like a small-town music festival, with people in lawn chairs, a high school brass band, scout troops selling food, and at least a hundred paddlers and their crews. The ten-minute pit stop we'd planned turned into forty-five. Christina was late because her navigation app took her to the wrong park in Waverly, one dedicated to the Confederate soldier Joseph Shelby. Shelby owned Waverly Steam Rope Company: a 700-acre plantation operated by enslaved people. In the Civil War, Shelby commanded a division of Confederate troops during General Sterling Price's Missouri Expedition, a failed Confederate advance across Missouri that ended in defeat at the Battle of Westport in Kansas City. After Lee's surrender at Appomattox, Shelby refused to lay down arms and so then led a group of soldiers to Mexico in June of 1865, where they tried unsuccessfully to join the army of Emperor Maximilian. After helping Price establish the ex-Confederate colony of Carlota in Mexico, he returned to Missouri for the rest of his life.

Thousands of Confederate monuments were erected in the South between 1890 and the 1930s. The Shelby monument in Waverly was dedicated in July 2009. In an era when Confederate statues were being dismantled, it's among the newest monuments to a Confederate leader in America.

We taped on red and green bow lamps and attached the white stern light with a copper clip and metal bracket, and I shoved off at 7:15 p.m. For now, the paddling pressure was off. I just had to make the sixty-eight miles to Glasgow by 4 p.m. the next day, and I expected to have more time to savor the landscape as I paddled into the heart of Little Dixie on Day 2.

The sun dipped below a line of trees. Crickets and cicadas droned. A chill set in, hard to believe after the afternoon swelter. For the first time in the race, I began to *see* the Missouri River. The channel curled between forests of sycamores and cottonwoods, with no bluffs or hills visible in the distance, only the river itself and, as night approached, not even the memory of wind. The moon rose. The conga line of boats reformed as an actual conga line— a floating party of weary paddlers. White stern lights decorated the path like strung Christmas lights. Kayaks floated broadside in the darkness—paddlers resting to eat sandwiches and apples and handfuls of trail mix pulled from cooler bags. A voyageur canoe breezed by with eight paddlers singing along with Queen: "We Are the Champions."

Don't get ahead of yourselves, I thought.

The sound of a live banjo came from shore at a tight turn near a massive tangle of cottonwood logs. It was the opening notes to "Dueling Banjos," the theme song from "Deliverance," a movie about four friends who take a canoeing trip together in Georgia that turns to horror when they are attacked by violent locals. Welcome to the backwoods! Some clever person had bungeed an mp3 player to a drift log.

For the next twenty-six miles to Miami, I learned to trust my ears more than my eyes in the black. Water pouring over wing dikes and around navigation buoys makes a rushing sound. Prior to the race, I hadn't practiced night paddling, thinking it was safer to learn in situ with paddlers and rescue boats nearby. MR340 veterans often describe the experience as transcendent, but I wasn't yet charmed. The lack of depth perception was unnerving, and the boat felt tippy in the darkness. Beyond Hills Island—a big sandbar—a loud scream came from near the bank. In the shine of my flashlight, an eastern screech owl took off from a wing dike carrying either a muskrat or a wood rat, a huge load for such a small owl.

When I reached Miami at 1 a.m., my GPS showed 105 miles for the day. Even in the middle of the night, volunteers waded into the river and brought me and my boat onto land, hauling the canoe up into a clutch of boats. I needed to sleep right away and set up the tent amid the boats instead of at the campsite farther off where some of the paddlers, or their ground crews, sounded like they were having too much fun. Two guys dropped a tandem boat within a foot of my head: "Wow, didn't notice you, bro." I set an alarm for 4:15 a.m. Somebody in the food line asked when the menu switched over from hot dogs to pancakes. A woman said, "Not 'til 3 a.m. honey. I don't make the rules, just the pancakes."

Driving south of Miami on Missouri Highway 41, it's difficult to imagine that the endless miles of patented corn were once a solid exuberance of tallgrass prairie. A month before the MR340, I had driven a route parallel to the river to search for unplowed grasslands that had managed to hang on in central Missouri. There weren't many. But old maps of the region contained hints about the region's preagrarian past. One, the Petite Saw Plains, was a spring-watered wet prairie seven miles long and three miles wide. Spared for a time by antebellum hemp planters who preferred dryer land, maize farmers felled the slough grass (or ripgut, as some call it because of the plant's lacerating blades) landmark after hemp farms shut down following the Civil War.

Another prairie on old maps was far more significant: the defining landscape of the Missouria people, who lived for centuries in the "big bend" of the Missouri near the mouth of the Grand River. The Pinnacles are a range of loess bluffs that rise dramatically from the flats south of Miami. An 1881 history of Saline County notes that the steep hills were covered by "one of the loveliest and most fertile prairies that ever charmed the eye of man."[5] Those native grasslands are gone today, grown up in forests or planted in corn and soybeans. Nearby Marshall has a city park honoring a psychic dog, but nothing that commemorates the sensile landscape the region was founded upon. Saline County is haunted by lost prairies, but one strong echo remains.

Michael Dickey is a retired historic site administrator with the Missouri Department of Natural Resources. Dickey's book, *The People at the River's Mouth: In Search of the Missouria Indians*, is one of the few book-length treatments of the people for whom the river and state are named.[6] Dickey told me that speakers of the Chiwere Siouan dialect originally lived in the Great Lakes region near Green Bay, Wisconsin. "Sometime before 1500," he said, "the Ioway, Missouria, Ho-Chunk, and Otoe broke into separate political entities and branched out."[7] The Ho-Chunk remained in the north. The Ioway settled in modern Iowa. The Missouria came to the big bend of the Missouri River and eventually to a village in the Pinnacles. The Otoe separated from them, settling near the mouth of the Platte River.

One of the first Europeans to see the Missouri River and learn of the Missouria people was Father Jacques Marquette, a Jesuit missionary who lived near present Duluth, Minnesota. Dickey said, "Marquette descended the Mississippi River with Louis Joliet and a group of Peoria guides in June of 1673." The confluence of the Missouri and the Mississippi was a turbulent one, and Marquette wrote in his journal: "I have seen nothing more dreadful. An accumulation of large and entire trees, branches, and floating islands,

. . . so impetuous that we could not without great danger risk passing through it. So great was the agitation that the water was very
muddy, and could not become clear."[8]

The Peoria guides told Marquette that the "mihsoori" or "wemih-
soori" people lived on the river they had just crossed. In the Peorias'
Algonquin language, the name—later anglicized as Missouria or
Missouri—meant "people of the wooden canoe." Canoes made of
wood were required for navigation on the unruly Missouri due to
its rapid current and many snags, while bark canoes were more
common on the Mississippi and rivers farther north. The Missouria
called themselves Niútachi, or People of the River's Mouth, referring to the Grand River.[9]

The Missouria civilization was centered around the Pinnacles,
and evidence of their presence has been discovered in at least
twenty-eight archaeology sites south of the river in modern Saline
County. In 1935, a farmer plowed up a 9" × 6½" Catlinite tablet that
Dickey told me was etched on one side with the rendering of an
ivory-billed woodpecker and on the other side with the head of a fal-
con and the constellation today known as Cassiopeia. Catlinite, the
sacred pipestone, is a type of indurated red clay, quarried from only
a handful of locations in Minnesota and Canada, and used by Native
Americans for centuries to carve pipes and other sculpted pieces.

The field where the tablet was found became known as the Utz
Archaeological Site, and further exploration revealed a major village, mounds built near the tops of hills, and a series of remarkable
earthworks known as the Old Fort. The ancient structure is one
continuous, irregularly shaped oval of ditches and embankments
enclosing six acres. Dickey told me that archaeologists believed it
was originally constructed by an earlier Woodland-era people, but
later excavations determined it was built by the Missouria. Despite
its name, there is no evidence it was ever used as a fort. Dickey believes it was likely a ceremonial area and celestial observatory.

Van Meter State Park contains most of the village site, the Old
Fort, and the burial mounds. The trail next to the Old Fort winds
through a mature oak forest, thick with groves of pawpaws. Although it is difficult to imagine the now-wooded Pinnacles as open

prairie, I could see how the village at the base of the hills, protected from the river, was in an ideal location, nestled between open grasslands to the east and the earthworks, burial mounds, and Missouri River to the west.

According to Dickey, archaeologists believe that, prior to European contact, as many as 10,000 Missouria occupied the village. By 1702, however, missionary Marc Bergier reported that the Missouria "are practically reduced to nothing."[10] An epidemic had wiped out most of the Nation and just eight hundred to a thousand people remained. They left the Pinnacles around 1720 and settled on Petit Osage Prairie about ten miles to the west. About 1792, an on-river ambush by their enemy the Sac and Fox nearly exterminated the Missouria.[11] The survivors abandoned their village. About thirty families joined their kinsmen, the Otoe, at the mouth of the Platte River. Another twenty families joined the Little Osage in southwest Missouri. Four or five families joined the Kaw people in central Kansas. In 1855, the Otoe-Missouria were forced onto a reservation on the Big Blue River near the state line of the future Nebraska and Kansas. In 1881, the Nation was again forced by the US government to move, this time to their current home near Red Rock, Oklahoma.

Today, the Otoe-Missouria Tribe of Oklahoma has over three thousand enrolled members and an active program to preserve the Chiwere language. The Missouri American Indian Cultural Center at Van Meter State Park is one of the only public museums or parks dedicated to Native American history in the tallgrass prairie region.

After I hiked the striking ramparts of the Old Fort, Dickey showed me one more map. At the northeast corner of the park, in the heart of the old village where the Utz Tablet was found, park managers have created a 40-acre tallgrass prairie restoration. A similar 80-acre restoration was underway at Arrow Rock State Park in eastern Saline County. Hiking through the new prairie at Van Meter, I could see a utility cut at the top of the bluffs where the river would have been visible two hundred years before. At least one stripe of prairie adorned the Pinnacles. A dickcissel landed on a post and made its onomatopoeic call. In the Midwest, dickcissels are quintessential prairie birds and, although they visit crop fields, it was evidence, perhaps, that prairie could one day return to historic grassland strongholds like Saline County, one of the most prairied counties of Missouri in pre-European settlement days.

When I stepped toward the bird, it flew off along the right of way and vanished over the clearing at the top of the ridge.

Only a fool paddles into the fog. This basic paddling axiom came in a Scott Mansker race dispatch email. Software like the MR340 Paddler app uses satellite-precision maps to video-game paddlers into the precise center of the river channel, helping them avoid run-ins with wing dikes, bridge peers, and sandbars that appear as animations on the screen. But mappable waypoints aren't the only dangerous stationary objects in a river. Navigational buoys, moored sand dredges, and moving barges do not show up on GPS-enabled apps. Barge pilots can sit above the fog line and navigate by radar. They would never even feel it if they struck a kayak or canoe, invisible to them in the murk.

At the Miami boat ramp the next morning at 5 a.m., the river seemed clear of fog, and three paddlers in surf ski kayaks launched into the flow. I threaded the camelback hose through the seat webbing, velcroed it to my life vest, and followed.

Fog came in wisps at first, flowing down the low banks of a soybean field, soon filling the river like a pitcher of fibrous vapor. I knew the rules and paddled cautiously toward the lee side of a wing dike. A man and woman in matching kayaks were eating crackers with brie and apples. They were playing peaceful music with a down-tempo vibe, the kind of music that couples who spend a lot of time with their cats listen to. The soft rhythms transformed the featureless fog into a ballroom; like floating at the edge of sleep toward an eerie infinity pool—one with brown water, beavers, and barges. They handed me crackers and cheese, and we held there in the shallows. After thirty minutes, a slight wind rose and the fog retreated back into the soybean field like a movie playing in reverse. The couple paddled off ahead of me, and I drafted them for as long as I could. After twenty minutes, I was alone for the first time in the race.

During the next thirty-six miles, the enormity of my predicament sank in. Scheme as I might, there was no way to make the race easy. The sprint to Waverly took everything I had, and there were still 236 miles to go. The next checkpoint was Glasgow, the 144-mile mark. From there, 200 miles remained, not even halfway to the finish.

Worrying thoughts gathered in my mind like storm clouds as I ferried around logjams at the mouth of a big river on the left. It was my old friend, the Grand. I caught the eddy line and paddled upstream a few hundred meters, then turned the boat back into the Missouri, officially "finishing" my float of the Grand, the river that ate my beautiful boat, the *Bluestem Rises*, and almost defeated me in my quest to know the rivers of the prairies. The Grand reminded me of my mission: *just keep paddling*, as MR340 superfan Chris Luedke repeats like birdsong in his online video series.[12] I decided to let the storm clouds boil up in my mind. I'd hunker down in the cellar of my solo canoe and keep going. In the MR340, mental grit is more important than anything else besides hydration.

I reached the boat ramp at Glasgow in low spirits. Day 1 was sprinting to Waverly. Day 2 was payback. Glasgow looked like a picturesque river village—I'm sure it is—but during the MR340, the town is a place where accounts get settled. Christina was waiting across from the city park. Two despondent racers were loading their kayaks onto cars while their ground crews stood looking away, arms crossed. Flies buzzed around a dead opossum on the side of the road.

We cleaned up the boat and Christina helped push me back onto the river. Below Glasgow, the Missouri follows a gentle but noticeable "downhill" gradient for one of the longest straights of the river. The woods were washed out in a shimmery sunflower haze and cicadas started to hum.

After two hours, the river broke right, and the race app urged me to follow it, rather than taking a side channel on river left. A sandbar at the head of a turn parlayed into a series of islands and me-

FIGURE 8.1 Lisbon Bottom, spawning site for the endangered pallid sturgeon (*Scaphirhynchus albus*). Big Muddy National Wildlife Refuge. Photo by Steve Hillebrand.

anders. The chute on the left sucked water from the river in a loud rush. I'd reached Lisbon Bottoms, a braided run that offers a rare glimpse of the preregulated Missouri River, the results of flooding and, ironically, careful hydrologic engineering.

After the flood of 1993—still considered the costliest in US history—levees caved in and the river poured into its bottoms, covering miles of cropland with several feet of sand and sediment. Miles of levees were totaled. The river forms a long sweeping S-curve through Lisbon Bottoms and Jameson Island, a broad flat near Arrow Rock that was no longer an island after the river had moved a century earlier. During the 1993 flood, the river tried to chisel through the S-curve to form a new channel, following natural chutes previous floods had fluted in the alluvial plain.

Jameson Island and Lisbon Bottoms were two of the first units of Big Muddy National Fish and Wildlife Refuge. At 16,000 acres and growing, the refuge is part of a big vision: to string together and restore a system of river bottom preserves that will let the Missouri River regain a facsimile of its historic dynamism and move about in parts of its floodplain.

According to the Big Muddy National Fish and Wildlife Refuge Comprehensive Conservation Plan, the US Fish and Wildlife Service had been looking for opportunities to increase protection for fish and wildlife along the lower Missouri since the 1970s.[13] After the flood, the agency obtained funding to acquire river bottom acreage from landowners who were tired of dealing with the river. Approved at first for 16,228 acres across eight Missouri counties, the agency authorized an expansion of the preserve to up to 60,000 acres between Kansas City and St. Louis in 1997. Together with the US Army Corps of Engineers, an agency that purchases riparian land for channel mitigation under the Water Resources Development Act of 1986, the projects are a symbiotic initiative to help flood-distressed landowners and begin to address the loss of almost 300,000 acres of wetlands, riparian forests, grasslands, and aquatic habitat after one hundred years of bank stabilization and channelization along the lower Missouri.

Tim Haller, a park ranger for the refuge, told me that the units of the preserve are like beads of a necklace. "First, we want to restore and maintain native habitat to support wildlife. Second, we want to promote biodiversity and allow for an abundance of endemic fish and wildlife species that were present within the Missouri River floodplain. Finally, we want to provide opportunities for people to enjoy the wildlife on the refuge and learn about the natural processes of the river."[14]

The flood at Lisbon Bottoms created a unique opportunity to restore side channels—a natural feature of big rivers like the Missouri—to a section of the river while still maintaining the navigation depth in the main stem. Haller said, "The Corps reengineered things to allow water to continue to flow through the two side channels at Lisbon and Jameson Island. This restored natural processes that hadn't occurred in 100 years."

In pre-European settlement times, low bottoms with expansive wet prairies of switchgrass, Canada wildrye, and prairie cordgrass grew in abandoned river channels and long flats. Haller said, "We don't have many remnant wet prairies, but we're reestablishing [them] on some of the units like Baltimore Bend, Overton South, and Berger Bend. But then again, we have periods of drought, and areas that have historically been wet are now drying out, and we're getting a whole change of vegetation. We've been planting prairie cordgrass in those historical areas. In places where we can burn and work against the influx of cottonwoods and other wind-dispersed seed trees, we've got a pretty nice response. But the challenge is conditions are always changing."

Because the preserve was founded on flood-prone areas, the river is connected to its floodplain in many of the units. Haller said that, in low water conditions, you can hike out to Lisbon Bottoms, or White's Island at Cranberry Bend, or Baltimore Bottoms, and feel something of the river's historical characteristics. "You can see it for what it once was." Or what it might become one day again, as the Big Muddy National Fish and Wildlife Refuge begins to realize its full potential.

Haller said that the refuge has been popular in Missouri. "When I talk with farmers, I can show that when our refuge floods, we take pressure off their land because that water is dispersed out onto our units and not pushing against their levies. Eighty-nine percent of our lands are available for hunting and fishing. Missouri is one of the only states in the country where voters passed a statewide tax for conservation. People in Missouri value the land and the heritage we have with it."

Lisbon Bottoms, especially, offers tantalizing hope for a species whose heritage predates not only the state of Missouri but the Missouri River itself. In 1998, US Fish and Wildlife personnel made a grail-like discovery when they collected larvae of the federally endangered pallid sturgeon (*Scaphirhynchus albus*) in a side channel there—the first non-hatchery-raised pallid sturgeon found in the Missouri River in over fifty years and proof that at least some fish were still spawning in the wild.

Pallid sturgeon are throwbacks to prehistoric times, little changed since the age of the dinosaurs. Endemic to the Missouri and lower Mississippi Rivers, it's astonishing that a creature as ancient and strange as the pallid sturgeon exists in a modern context. Pallids, as fisheries professionals call them, can live for a hundred years, grow up to six feet in length, and weigh as much as eighty pounds. Instead of bones, their body is girded with embedded cartilaginous plates, and they lack scales. The species is aptly named "pale," with some appearing colorless and others a light gray. Nearly blind, they navigate mucky river bottoms using specially adapted whiskers and barbells. Fewer than two hundred are thought to survive in the upper reaches of the Missouri. As many as 21,000 still exist in the lower reaches of their habitat.[15]

Despite surviving the extinction of the dinosaurs, the birth of the Rocky Mountains, the evolution of the Missouri River system, and multiple glacial epochs, pallid sturgeon have been driven to the brink of extinction in the last century because of changes to the Missouri River itself. They were designated as a federally endangered species in 1990.[16]

Wayne Nelson-Stastny, a biologist with the Missouri River Recovery office of the US Fish and Wildlife Service, described the pallid sturgeon as nomadic. "This is a fish that needs long stretches of river. They come and go within the system, taking what they need at certain seasons of the year and of their lives. The adults will travel hundreds, if not thousands of miles, if given the chance."[17] When a fish bypass was added to a dam on the Yellowstone River, he said that "they almost immediately traveled 157 miles up from the dam into Wyoming, a place where we hadn't seen the species before. It happened very fast."

The principal cause of their decline appears to be a lack of recruitment. In the wild, virtually 100 percent of embryonic sturgeon die within their first year of life, hindering the ability to replenish populations.

In the upper Missouri, large reservoirs pose the greatest challenges. Bob Bramblett, the biologist who studied fish assemblages in the Purgatoire, has also worked with pallid sturgeon in Montana. He told me, "The big reservoirs like Fort Peck have had a huge impact, it's the real game changer for river ecology. What's important is where the dam is located on the river continuum between the headwaters and the mouth."[18] Mainstem dams often release cold, clear water from the lake bottom into the river below in a process known as hypolimnetic release. Bramblett said, "Fort Peck is strongly out in the prairie. This is a warm water ecology. Pallid sturgeon are adapted to warm, muddy rivers, but now, the Missouri has stretches below reservoirs with colder, clearer, faster moving waters."

When sturgeon embryos drift into reservoirs on the upper Missouri, biologists believe they sink to the depths and perish in cold, oxygen-depleted environments. "Once the embryos drift into the reservoir, that's it," Bramblett said. "They don't recruit. They're not able to reproduce. The alterations are cumulative. Today, you've got a fragmented river that blocks the movement of the adults. Upriver you have the reservoirs that create clear cold water and alter flood regimes. We think the reservoirs kill all of the embryos, although, since it's so hard to find them in these deep reservoirs, it remains a hypothesis."

In the lower Missouri, channelization has discouraged the formation of side channels and sloughs connected to the river. According to Nelson-Stastny, "as embryos, once [pallids] hatch, and before their fin folds develop to the point where they can hold their place in the stream, they begin to drift in the current. In a natural river system, they might drift for two hundred or three hundred miles and wind up in a shallow side channel or meander. But today, if a sturgeon spawns in Yankton, South Dakota, it could be nine hundred miles away in the Mississippi River in less than a week."

Flood years provide evidence to support the theory. During high-water events, the river can breach levee systems, flooding adjacent forests and fields and creating temporary side channels, as was common in pre-European settlement times. Nelson-Stastny said that Nebraska state fish crews studying the related shovelnose sturgeon found flooded areas to be "highly productive sturgeon nurseries. Young fish could spend the first parts of their lives there, eating mayflies and insects emerging from the bottoms where organic material and substrate pops up. All sturgeon species are incredibly adapted to turbid muddy waters. Their barbells and mouths are like vacuum cleaners. They search the bottoms with their barbells and can suck up and sift through the muck to pluck out insects." The studies revealed that flood years resulted in thriving populations of "plump" sturgeon. "These floods create tough years for humans, but for sturgeon, we see more reproduction, more fish, more embryos drifting."

Today, young pallid sturgeon are reared at facilities like Gavins Point National Fish Hatchery in Yankton, South Dakota. Thousands of tagged pallids are released into the Missouri each year in efforts to stabilize their populations. Given the bleak prospects for the fish in the upper river, restoring slow channels and sloping bank habitats in the lower Missouri—at spots like Lisbon Bottoms and other units in the Big Muddy NWR—could create critical sanctuaries for these fish. There sturgeon embryos might have a chance to survive long enough to perpetuate a species that has thrived in this river for millions of summers.

It's 3:30 p.m. and 97 degrees. Thermostat maxed. Pulse, 150 buffalo hooves per minute without paddling . . . 170 miles behind us, 170 miles remain.

After a brief refueling stop in Boonville, I paddled fast away from the ramp, bearing down when malevolent currents at the end of an island spun me sideways. The hard work cleared my head and sinuses. The forest beside the river smelled of wet earth and leaves. A lightness filled my body and, for no reason, I felt almost giddy zipping downriver away from the setting sun. For the next ninety minutes, I kept up the exuberant pace, six strokes on the left, eight on the right. I turned on the race app with its gamer green animations of the channel, dropped my phone into a waterproof bag, and cracked a red glowstick to set next to the phone.

My reverie was brief. As night latched down upon the river, I found myself alone again. The smartphone navigation charts got me safely to Rocheport, where wooded bluffs congregated on river left. Beneath the I-70 bridge—a span I'd driven across countless times, curious about what the miles of wetlands would be like in a canoe—I tightened my grip on the paddle. Algebraic currents caused by the huge girder beams almost twirled me, the gnarliest water of the race. Later that night, a paddler flipped in a whirlpool, "the kind you can hear." He managed to grab his kayak and reach shore with the help of another paddler, but sank almost to his waist in mud and quicksand while hundreds of mosquitoes bit his face.

I had never felt so alone on a river, and not in a liberating way. When two green lights appeared, I paddled toward them. Two brothers, moored in a canoe off channel, were looking at their phones. "Damn, I'm glad to see you guys," I said.

They planned to push on all night if they could wake up and find their pace. We waited until another solo kayak showed up and, without exchanging names, paddled off into the northern lip of the Ozarks, where the river began carving through what remained of the ancient mountain range.

The next hour was magic.

We formed a tight drivetrain of boats and followed the channel like a footpath. The hills rose first on river left, then on river right, a palisade of actual rock, not of loess soils like the bluffs back in Kansas City and at the Pinnacles. Pushed by the current toward a cliff that towered at least 150 feet, we shone our flashlights against its sheer high-rise.

A neolithic moon emerged on the brink of fullness and began pouring out like honey over the Ozark bluffs in a scene too beautiful to ever describe with words. Bats flickered across the water in terse patterns, skimming mosquitoes and katydids. Swarms of no-see-ums stung our faces, but nobody complained; we brushed them away, drowned dead in sweat, and continued paddling beside the black hills.

Moonlight permeated the empty cargo hulls of our boats and illuminated paddles, clothing, and dry bags like a twinkling linen. One of the guys flashed on his headlamp for an instant, and sparrow-sized luna moths circled like a halo of forest nymphs. Where the green starboard lights shone through the top two feet of river, prehistoric gar with crocodilian teeth rose from the murk, attracted to shiners that swam toward the light. Time seemed to stop as our boats continued along with the current in total silence. We were four regular guys with day jobs, less than two miles from one of the busiest interstate highways in America, astounded by sights that few people had ever seen. We broke the soft water with our paddles only when we had to, taking great care not to disturb the dreamscape on this greatest of grassland rivers.

At midnight, a boat ramp volunteer grabbed my hand and said, "Welcome to Cooper's Landing." I'd paddled ninety-three miles from Miami, the second longest paddling day of my life after the 105 miles of the day before. My legs were concrete stilts walking to the top of the ramp.

Then, in a freakout of confusion and burning rubber, a pickup truck that was parked on the street in front of me gunned its engine, tires liquefying in place, and lurched forward over the curb, banging into a tree and a parked car, heading straight at me. In an adrenaline freeze frame, the car stopped one canoe's length short of my certain maiming. I looked the driver dead in the eyes. He put it in reverse, struck another crunching blow into the parked car, and peeled out onto the dark road that paralleled the Katy Trail bike path.

I stood in shock, holding my breath. What were the odds of getting run over by a crazy driver at the top of a remote boat ramp in the middle of the night? It happened so fast—and I was so river beat—that I wouldn't have been able to dive out of the way. My mother asked only one thing of me in the race: *Don't get killed.* She should have been more specific.

As I climbed into the camper—Christina had arrived before and parked next to the tent camp—she asked, "Did you see that truck get pulled over by those guys?"

We could hear loud voices and shouting. "I'll pay to fix your car." Then, flashing lights.

Day 3. 4:15 a.m. Text message from the race: "Local Barges blasting up and down the river between MM144 and MM139."

Cooper's Landing was a mass of filthy paddlers, sleeping like hi-

bernating animals on asphalt, and in tents, cars, and campers. I
paddled into the darkness with rigor, the hard paddling caffeine to
my groggy sinews.

Crows were waking up in the woods along the river. Their calls
were not the "kaw kaw" of the American crow but the nasally kents
of the fish crow, *Corvus ossifragus*. From Cooper's Landing down,
most of the crows were fish crows, a southern bird that prefers tide-
waters, swampy wetlands, and the narrow corridors of large rivers and their tributaries. Fish crows have moved west into prairie
states in recent years, probably expanding their range as the climate warms. When one flew close and made its asthmatic greeting,
I yelled back, "I hear you!"

A woman in a camo-kayak paddled by. "Talking to crows?"

The previous day, I'd averaged close to six mph, but now I was
down to a laconic 3.6 mph—slower than the Kaw—with a hot head-
wind and no other boats around. At this pace, I'd reach Hermann
in the middle of the last night of the race. Moving at the clip of a
toy boat in a bathtub and my heart rate rising with the noon sun,
I nearly flipped the canoe when three long, sonorous, fog-horn-
sounding blasts wailed from behind me. The barge blew its horn
again, signaling me to move off channel. With manic agency, I paddled orthogonal to the current and cruised into a small tributary
creek to let the gigantic boat pass.

The Mississippi River system, or inland waterway, is a 12,000-
mile transportation corridor where goods and bulk cargo move
along rivers that connect to the Gulf Intracoastal Waterway at New
Orleans. Besides the Mississippi itself, barges traverse parts of the
Ohio, Arkansas, Illinois, Tennessee, and Missouri Rivers.

In 1981, the US Army Corp of Engineers completed the 735-mile
navigation channel on the lower Missouri, authorized by the Rivers and Harbors Act of 1945. By then, commercial barge traffic had
begun a slow thirty-year decline from a peak of 3.3 million tons of
cargo in 1977. A 2003 study reported that grain shipments on the
Missouri—one of the primary drivers for the navigation channel in
the first place—fell by 81 percent from 1964 to 2000, and wheat shipments fell even faster, from 1.77 million tons in 1964 to 21,000 tons

in 2000, a 99 percent drop. The study called out the long distances between ports on the Missouri that translated into higher fuel costs for towboats, local ethanol production from plants that rely on trucking, declining US corn and wheat exports, low flows from droughts and reduced releases from upstream reservoirs, as well as competition from railroads. According to the study, in 2001, "one railroad alone hauled approximately the same number of bushels of grain from the five states bordering the Missouri River to Mississippi River barge markets at St. Louis as barges hauled on the entire Missouri River in 2000."[19] The report concluded that an open public debate was needed to consider whether public investments were worthwhile on low-volume rivers such as the Missouri.

Since the report was published, some of those trends have reversed. Dramatic declines in barge traffic started to shift in 2007 as the state of Missouri and private companies began investing in new ports along the river. The Missouri port authorities were chartered in the 1970s to support economic development of Missouri's waterways—primarily the Missouri River. In 2024, Missouri had fourteen ports, with major operations in St. Louis, Brunswick, and Kansas City. Since the nadir of barge traffic in the 1990s, the state has built a sophisticated network of intermodal facilities that connect barges to rail and truck transport.

Barge captains, crews, and port workers who depend on the river for their livelihoods know the Missouri as well as anybody. An intimate knowledge of the river's geography and temperament is essential to navigate two-hundred-by-thirty-two-foot rigs that draw a maximum of eight and a half feet of water. A typical barge can carry fifteen hundred tons of freight, the equivalent of about sixty tractor trailers or eighteen jumbo rail cars.

Richard Grenville is responsible for port terminal operations and business development at the Kansas City Port Authority. A London native, Grenville got his start with the British Merchant Marine before coming ashore in Savannah, Georgia. Eventually he traded blue water for brown when he took up positions for the Port of Tulsa on the Arkansas River and later the Port Authority of Kansas City.

Grenville told me that barge transport is trending up on the Missouri. He believes that environmental concerns and traffic safety could be behind some of the increase. "You can put sixty truck trailer loads on a barge, and a normal tow on the river pushes eight barges up and six barges down. That saves 400 trucks and the fuel they consume. It's also a lot safer. There are less accidents on the waterway than on the roads or rail. And you're reducing emissions because there is only one barge tow pushing the entire load."[20]

Barges on the Missouri transport products for agriculture, minerals, and grains produced close to the river. The Kansas City Port deals mainly in fertilizer and grains but also commodities like mill scale, a byproduct of steel production. Grenville said, "When the steel comes out of the steel mill, it has all this flaky stuff on there like a rust. It's iron oxide, and they used to bury that. Now cement companies use it to mix in with their cement. It makes a very hard product. We also deal with coal slag, the unburned minerals from coal. They're used as abrasives. That black sandpaper you get at Home Depot, that's actually coal slag on there. It comes from the bootheel of Missouri, from a power plant, and we send it to La Cygne, Kansas, to a company that produces abrasives."

Some 3.6 million tons of sand and gravel were shipped on the river in 2022, as well as a million tons of fertilizer, grains, scrap, steel, minerals, and petroleum products. That's triple the volume from 2015. However, even though the Missouri Department of Transportation reports that barge traffic has been going up for a decade, it still pales when compared to the Mississippi and Arkansas Rivers.

The tow that honked me into the little tributary pushed four barges of rock, a crushing load. The rig sagged low in the current. Barges can't stop. Because the only place in the river barges can navigate is in the deepest water, paddlers must get off the channel when one approaches. The channel is marked with steel buoys and signs attached to cottonwood trees and rock levees placed at crossing points, a constant reminder that the river is an interstate highway for barges. It is important to either get off the river or position your boat at a slight angle to the barge wake to avoid being flipped. For as long as one hour after a barge passes, the water is choppy, like a hot tub with the jets turned on.

In an early running of the MR340, a couple from California was run over when they mistook a barge for a mired sand dredge. They were lucky—the force of the wake pushed them down and up on the sides of the big rig. They emerged unhurt, but the tow's propellers chopped their kayak to bits.

The creek I'd paddled up was foul-smelling and soapy, but a fine refuge. There was a rustling on the bank, and a beaver the size of a small pig crawled up onto a cottonwood log and watched me while munching on a willow shoot with orange buck teeth.

In Hermann—the river town settled by German immigrants who founded their "Missouri Rhineland"—Christina met me with an entire pizza. This cemented her celebrity with the other paddlers who shared it.

Soon, I was back on the river headed toward New Haven, seventeen miles downstream. The last racers at Hermann were already gone, so I was on my own for this final night of the race.

When the last glint of twilight faded, I heard a sound like rushing water. It got louder, and eventually it seemed like I was surrounded by waterfalls. In Hermann, we had studied the maps and I thought that a tricky spot on the river, Berger Bend, was beyond New Haven. I'd misread the maps. A large rock spire splits the river at the bend. At night, or in fog, paddlers must avoid the chute on river left at the end of the island. In previous years, the chute had turned racers around. Some spent maddening hours paddling in circles around the rock. I knew to stay in the middle of the river and relax; there are no waterfalls on the Missouri River. The channel, however, was booby-trapped with a platoon of bouncing green buoys, some, no doubt, pinned down by enormous trees pushing on their cables. This added to the rushing noise. I gave each buoy a wide berth, especially since the river was swift and tried to pull me toward them. Grinning whirlpools did dervish dances below each buoy.

Despite the moonlight, the forests were sloping black pools of darkness. I hadn't memorized the river mile of my stop for the night and wasn't sure how I'd find New Haven. For no good reason besides entombing fatigue, I began to hula-hoop that uncertainty around in my mind. The river had narrowed and deepened. My paddle slicked through black boils six feet across, rising like smoke rings from the depths in the fastest water of the race.

Things got weird. Suddenly, it seemed perfectly rational to stand up in the boat and walk toward shore. The river looked like black-top, asphalt. I knew the thought was madness, but it kept clawing at me.

Faces came next, monkey faces, poking out between branches of trees in the dense Ozark woods. The trees became menacing. Walking calmly from the canoe across the riverine parking lot seemed like beautiful peace.

What was in that pizza?

Then I remembered where I was. Between Berger Bend and Washington, MR340 paddlers have reported hallucinations, often shared, coming from the forest. I was in that section of river. The visions were part of the race, apparitions born of voluntary sleep deprivation, a lack of REM cycles, and golden moonlight filtering between branches. I had been sleeping three to four hours per night, so I wasn't a total zombie, but I was zoned enough to see monkeys in Missouri trees. I switched paddles to shock my muscles with a heavier load.

Finally, a light appeared on river right, then a row of lights, and then the sound of voices, real people's voices, not monkeys. Still unsure how to land, I called Christina. She said, "Go past the jetty, sweep hard to the right, and turn into the eddy." The current nearly swept me past the ramp. Once I got past the jetty, I almost toppled the boat maneuvering the turn.

My plan was to sleep for four hours and leave before sunrise. I brushed my teeth and cleaned up in the pitch black of a cinder block bathroom with the lightbulb burned out. A train whistle blew, and when it got close, its light blasted through every crack in the walls, daylighting the inside of the bathroom.

On the last morning, the river was running fast, a mercy. Approaching St. Charles, the Missouri had entered a region once called the Grand Prairie. This close to the Mississippi, though, little actual prairie remained. Surveys of early St. Louis conducted by Antoine Soulard show that timber coddled the river near Laclede's Landing, but most of the future city was prairie and mixed prairie/savanna. French settlers maintained some of the prairies—with names like Prairie de St. Louis, La Grand Prairie, Prairie des Noyers, Prairie a Catalan, La Prairie du Cul de Sac, Prairie of the White Ox, and Prairie of the Three Bulls—as grazing commons. Today, those prairies are gone.[21]

So is almost all the original tallgrass prairie east of the Mississippi River. Future Illinois, once nicknamed "the prairie state," had twenty-two million acres of prairie. By 1978, only 2,500 acres remained. Prairies covered 15 percent, or about two million acres, of future Indiana. Only a handful of never-plowed remnants in the western part of the state remain, totaling about one thousand acres.

As America pushed west, manifest destiny devoured the prairie in its wake: via conversion to agriculture, cessation of cultural fires after the forced removal of Native Americans from their ancestral homelands, extirpation of grazers such as bison and elk.

The urge to re-create prairies *as ecosystems* began where they first vanished, in the eastern prairie states. Before the Conservation Reserve Program was created by the Department of Agriculture in the 1950s (the program pays farmers to take croplands out of production to replant native grasses—although not *prairies*) and before the federal government began to reseed nonnative crested wheatgrass on Dust Bowl–ravished farms purchased late in the Depression, the University of Wisconsin acquired 500 acres of farmland outside of Madison. Aldo Leopold, whose titanic 1949 work *A Sand County Almanac* pioneered ecocentric concepts in wildland management and made him a rock star in the environmental movement, had joined the university in 1933.[22] During his first winter as arboretum research director, Leopold and biologist Norman Fassett were given control of the 500 acres and decided to restore the native tallgrass prairie, an ecosystem that had covered parts of future Madison prior to European settlement. According to science writer Christina Mlot, Leopold and Fassett's prairie restoration may be "the oldest ecological restoration of any kind" in the world.[23]

Two years after students set up experimental plots and reseeded the basic grasses and a few species of forbs in 1933, a crew of 150 Civilian Conservation Corp rewilders set up a camp near the university and spent the next weeks digging plants from small donor tracts in cemeteries and along railroad rights of way to transplant. In addition to seventy-two acres of prairie, the rewilding crew also replanted transects of boreal and deciduous pine forest. The first stage of the restoration was completed in four years under the supervision of ecologist Ted Sperry.[24]

The Curtis Prairie, named in 1962 in honor of John Curtis, who conducted early research on plant community restoration at the site, is approaching its centennial anniversary. Graduate students have earned master's degrees and PhDs scrutinizing the site's soil characteristics, microbiota, insect and bird life, and the spatial relationships and distribution of native and nonnative plants to see how management techniques and time have affected the first man-made grassland designed as *ecosystem*.

Prairie restoration is a new yet vigorous science with dedicated journals and annual conferences. Botanists, biologists, landscape engineers, farmers, cultural anthropologists, philosophers, and ethicists often huddle together to chisel away at the hardest problems. What are the important goals of prairie restoration at both the micro and macro levels? How should *success* be measured? And, perhaps most important of all, is it even possible to reconstruct a prairie, or are the best restorations merely clever but doomed counterfeits that will never function like the originals? Recreating something as biologically complex as never-plowed tallgrass, mixed-grass, or shortgrass prairie is almost as complicated as reverse-engineering consciousness. Most researchers seem to agree that *perfection* is not the goal, but a wide range of opinions exist as to what is "close enough."

Ecologist, ethnobotanist, and environmental studies professor Kelly Kindscher has dedicated his career to studying tallgrass prairies and pondering these sorts of questions. His work has inspired a generation of students who work today at the forefront of grassland preservation and restoration. Kindscher once walked from Kansas City to Denver, keeping to prairie trails where possible and eating wild foods he found along the way. In a seminal 1998 study of two native prairies and adjacent restorations in northeast Kansas, Kindscher and his colleagues concluded that reaching biodiversity equivalence in restored and native prairies will take centuries if it can be done at all.[25] But "complete" restoration might not be an accurate or fair measurement, given that all ecosystems, native or restored, are dynamic systems in constant flux. One critical role of prairie restoration is to encourage unplowed prairies to expand onto adjacent lands or, better, reconnect discontinuous ones to build larger prairies.

Kindscher wanted to understand how that process works, how native prairie "invades" adjacent replanted prairies. He wanted to measure how native prairie spreads over time through "seed rain" from gravity and wind dispersal, or whether the original seed bank could be recruited by the restorations (some seeds remain viable for decades or longer after a prairie is plowed). Obviously, it wouldn't spread if the "neighbor" was a crop field that got plowed every season, but few studies had measured to what extent virgin prairies enhance the genetics and biodiversity of replanted ones.

Kindscher and his colleagues catalogued two restored prairies five and thirty-five years after their establishment. They discovered that not only did species richness decline with proximity to native prairie, a phenomenon known as the distance effect, but that difficult-to-establish species—most critically nitrogen-fixing legumes such as cream false indigo—were rare in restorations. Short of more aggressive interventions, restoring grasslands adjacent to existing prairies could take centuries, at least if success meant passing the prairie Turing test, or, as Kindscher calls it, *complete* restoration. Kindscher told me, "Not every organism needs all the pieces, some just need grassy habitat." But other species are more particular. "A discovery in the last ten years showed there's a moth that only feeds on the stems of *Eryngium yuccifolium*, or rattlesnake master. That's a species that has a relationship with one individual plant, and many restorations don't have rattlesnake master. It's in native Kansas prairies, but the seed isn't readily available. You can't buy it commercially."[26]

Like Kindscher, Chris Helzer, director of science for the Nature Conservancy in Nebraska and author of *The Ecology and Management of Prairies in the Central United States*, doesn't know if complete restoration is possible.[27] Helzer agrees that the most important goals are to take the pressure off remaining native prairies and to help them expand. He told me, "There's a trap we sometimes fall into, that we're trying to replace what's gone, by replicating what it looked like as much as we can. Small, isolated patches of native prairie are really vulnerable. It's tough to conserve their biodiversity. Eventually, in a small, isolated patch of prairie, populations of everything can trend toward local extinction. If we can make those prairies bigger and, even better, reconnect them with other sites so that species can move around, then we're a lot less likely to see the loss of an entire population of a species."[28]

Helzer said that managing restoration for biodiversity has hidden pitfalls. "If you burn the entire site, anything that's vulnerable to fire at the time you burn is probably going to die. In the dormant season, invertebrates and other animals that are living above ground are all going to get burned up. If you burn during the growing season, there's plenty of other things that are going to get damaged. There's really no safe time to burn if you're managing only for biodiversity. Haying can also be troubling; grazing can change habitat structure in a way that can eliminate species. So, if you do the same thing across the whole site, you're going to lose something. And at a small site, you get to the point where it's not logistically possible to keep splitting the place up into smaller pieces."

Helzer, like Kindscher, wants to focus on the irreplaceable native prairie that remains, not on perfecting restorations, and find better and more predictive tests to monitor the health of grasslands and the biological communities that depend on them. "The analogy I use is that it's like the difference between trying to restore a historic building or a car when you know what the end result should look like, and you can see how close you get to it, versus restoring a city after a disaster. There, you don't care whether a building goes in the same place or looks the same, or the communication tower is built so it looks just like the last one." The important thing, Helzer points out, is to get communication networks, grocery stores, roads, and power lines back online—to emphasize restoration of function over form. He said, "I think we should stop worrying so much about how a restored prairie looks, or what the plant composition is. We need to use native plants from local sources. We want as much bio-diversity as we can get. But we should look at whether the species in those remnants can breed successfully in the restored habitat. If that's true, then we think we're doing our job. In Nebraska, the news is good. Bees, ants, leafhoppers, fungi, grasshoppers, and small mammals all are moving out of these remnant prairies into restored prairies. We're finding very few, if any, species that can't do that. These species will be better off with larger, more connected populations. We've just scratched the surface."

Helzer's vision is to protect the last remnant of the primordial prairie ecosystem with huge buffers of reconstructed prairies. Helzer has seen this process pay off, both through his work with the Nature Conservancy and closer to home. "We've got a family prairie with a quarter section of land that was mostly farmed when my grandpa bought it in the late 1950s. He put the majority of what was left back to grass, just grass. And there were these little pockets, less than an acre most of the time, of remnant prairie. And so there are species of plants that we've seen moving up the hills and onto the formerly farmed areas. Some of them have filled in everywhere."

Chris Newbold, who has worked with the MDC as a natural history biologist and district supervisor, has seen this kind of successful restoration in a prairie close to the Missouri River. The Grand Prairie was a claypan prairie that once stretched west from

St. Charles almost to Columbia. One remnant that survives is the 140-acre Tucker Prairie, owned by the University of Missouri. Newbold's team was tasked with using seed collected at Tucker Prairie to create a new prairie on bare farmland near the Missouri River outside St. Charles. They took the field down to "bare canvas" by plowing the site for three successive seasons and applying herbicides that killed everything that grew. Then they planted the highly diverse seed mix hand-harvested from Tucker Prairie. Newbold said, "It was successful. We've seen 152 species show up, although there's still 100 species or so that we collected that we didn't see. We're trying small experiments to make it better, like using mycorrhizal adjuncts to see if that increases germination."[29] Newbold, like other researchers, believes new management techniques or scientific breakthroughs will be necessary to successfully reintroduce fragile species that are seldom observed in restorations.

What, then, is the best way to restore a prairie? I asked Courtney Masterson, founder of the nonprofit organization Native Lands, one of a small but growing number of prairie restoration operations in the heartland. Masterson said the process is different for each prairie, but restorationist pros like herself follow some version of the following steps.

STEP #1 CASE THE SITE.

"The first thing I do," Masterson said, "is observe the site to see if there are any weeds to manage. We want a blank slate, and the best way to achieve it is to delay planting long enough to observe what's going on."[30]

How long? I asked her. "It sounds crazy, but we would watch it a year if possible, especially on old farmland that could have herbicide-resistant weeds, or any other problematic things, in the soil. Once you put down several thousand dollars' worth of seed, it becomes very difficult to manage invasive species that you would have known about if you'd gone slow."

Another advantage of observing an entire growing cycle, Masterson said, is to look for micro-habitats in the planting area—wet, dry, rocky, sunny, or shady spots. "In a perfect world, you would customize your seed blend for each microhabitat. If you have any woodland or established trees nearby, there should be different blends for different micro-habitats in your planting area," she said.

STEP #2 PREPARE THE SITE.

Prairies were once plowed to grow crops. These days, Masterson said, restorationists plow crops to grow prairie. The closer to bare soil the better. Herbicides like glyphosate are used to further attack the weed base so native seeds have room to grow.

STEP #3 ACQUIRE (THE RIGHT) SEEDS.

Once the ground is prepared, the next step is to seed the new prairie. Choose seeds from species found in nearby native prairies. The more species the better. Masterson explained that many restorations use less than thirty species, and that can be good, but restorations with the seeds of more than two hundred species are becoming more common. "Look for seeds from local vendors, as nearby as you can, and make sure they are acquiring from local sources. When you buy seeds with genetics from totally different climates, your seeds are more likely to fail. We do a lot of hand collecting from remnants near our sites to stay hyper local, but also to maximize biodiversity. Some seeds are not commercially available. We go after those."

When they get ready to plant, Masterson mixes seeds in with soil or sand. That's the time to add mycorrhizal adjuncts.

"Especially on farmed land, you've lost almost all the fungi. Adding it back helps in general, but especially with prairie legumes dependent on mycorrhizal fungi for survival beyond their first year."

The success of legumes can mean the difference between ending up with a field with only a few dominant grasses and a diverse prairie doppelgänger. "Legumes improve the soil quality over time and, at an ecological level, provide valuable feed for wildlife. Grassland birds really depend on legumes."

STEP #4 PLANT THE SITE.

Because, unlike garden cultivars, most prairie species require cold stratification to germinate, seeds should be planted in winter. Masterson said, "Most seeds need two to three months of cold and wet to break their dormancy. So, we plant in the winter. It can be a great community activity. You get to break a sweat when it's too cold outside to do much else." In most of her restorations, landowners are right there with her to walk the land and sow seed by hand.

STEP #5. MOW AND ADDRESS WEEDS IN THE FIRST YEAR.

After seeds start sprouting, mowing is the most common management practice during the first year or two. Masterson said that mowing once or twice per summer addresses common weeds like foxtail, horseweed, and Queen Anne's lace.

STEP #6. SUPPLEMENT.

Starting the second year, restorationists boost biodiversity by planting plugs and greenhouse-reared plants of species that are almost impossible to start from seed. Masterson said, "Lead plant, New Jersey tea, rattlesnake master, and some of the orchids and milkweeds almost never come up from plantings on raw ground, so we add plugs the first or second fall, then the next season the plants bloom and start dropping seeds on the site, and it goes from there."

STEP #7 BURN, GRAZE, MOW, REPEAT.

In the second year, Masterson establishes a burn cycle. "We start burns in the second or third winter. That's just to pull some of that dead plant material off so we're not smothering out young native plants and to give them space to fill in. Then you're on a burn rotation from that point."

Borrowing a buffalo can help too. "Prairies are a disturbance-based ecosystem, so we have to develop a long-term plan for burning, mowing, and grazing depending on what the landowner has available."

STEP #8 TAKE IT EASY.

The final step, according to Masterson, is the most important for both restorationists and property owners. Be patient. Check each season to see what species are missing and add them with targeted plantings. Fight invasives. Continue to monitor. With the latest techniques, there is every reason to expect that prairies replanted today will significantly surpass the biodiversity and ecological functioning achieved in earlier restorations.

Chris Helzer is optimistic about the future of prairie in the Midwest. "We've learned that beyond a doubt we can establish highly diverse plant communities by using diverse seed mixes. We're using 230 species in our mix. If you were going to buy that, it would be unholy expensive, but if we could solve the price problem and if enough seed was available, the good news is we know that we can reestablish prairies that will persist for long periods of time. We've got really good examples across the Midwest and Great Plains of sites that hold onto their diversity."

Helzer finds reason to believe that prairie ecosystems might be more adept at long-term survival than previously thought, citing the rebound of grasslands after the decade-long drought of the 1930s. "During the 1930s, tallgrass prairies became basically unrecognizable to people like John Weaver, who was studying them back at the time." Weaver was known for his research on the root systems of prairie plants. "Tallgrass prairies became monocultures of western wheatgrass," Helzer told me. "Then eight years later, they were diverse tallgrass prairies again. They can adapt and change based on conditions. And the reason they can do that is because they have a lot of diversity, not just of plants, but of everything, so that there's always a species that's going to be well adapted to whatever is happening at the site. If you have that bench strength, those species in waiting that are at the site, as long as you have enough scale that those populations can maintain themselves, grasslands are incredibly resilient."

That could be important in coming decades as climate change plays out across the Great Plains and Midwest. Kindscher told me, "One of the few solutions we have for reversing climate change is to store carbon in soils." This is because soils act like a natural carbon dioxide sponge, removing carbon from the atmosphere and storing it underground. He said, "One of the best ways to do that is to plant more prairie. I think we'll see more initiatives to reconstruct prairies for this purpose."

If Kindscher has his wishes, this could dovetail with emerging technologies. "In the Midwest, it comes back to agriculture. My optimism is this: ethanol subsidies will probably end. We are not going to need that much ethanol if we're switching to electric vehicles and if we're going to remove a third of the corn crop acreage from producing corn, because that's what ethanol is. Instead, we could pay those farmers to grow carbon. A good way to do that is to replant prairies, because their deep branching root systems can hold onto significant quantities of soil organic carbon."

If programs like that emerge in coming years, they could provide farmers with new revenue sources to remain on the land, tackle the most significant challenge facing the world in the twenty-first century, and simultaneously promote biodiversity and the resurgence of prairies in the heartland.

Approaching the final bridge of the MR340, my joints were screaming. It was impossible to get comfortable. But it didn't matter. End-of-the-race landmarks rolled by like a movie—the "last turn," the outer fringes of urban St. Louis, the Missouri Highway 364 and Daniel Boone bridges. Boats passed me while I savored the final miles of misery.

During my search to know America's grassland rivers, I lived among wild rivers and prairies, and their wildness lived in me.

I often dreamed of wind swirling geometries into flowing grasses, of seed rain impregnating golden fields with the ancient genetics of primordial meadows, of brown waters parting respectfully around red quartzite boulders, of ivory shorebirds landing in unison like ballet, of lightning storms and rattlesnakes on the upper Missouri, of untamable floods pressing the Grand River to reclaim its sinuous heritage, of ripgut sluices in the Wakarusa Wetlands purifying the rainfall of a thousand thunderstorms, of lost souls reaching home after parched centuries on the Purgatoire, of riding the South Platte meltwater and the Niobrara Ogallala, of moonlit visions on the lower Missouri and, always, of slow green waters lilting and lolling my canoe—and spirit—downriver toward the next campsite. Unique in the pantheon of American waters, grassland rivers are fertile corridors for conservation, making personal connections to the land, and rediscovering the meaning of "discovery" itself. Anybody who seeks them will be wilder for the attempt.

I clutched my paddle and ferried into the heart of the big river, in the fast water. When the Lewis and Clark Boat House came into view, I cut across the channel in one deft motion, like I knew what I was doing.

Seventy-nine hours and twenty-two minutes out from Kaw Point, the boat thudded to a stop in soft mud. Christina helped me out of the canoe, and we walked up the ramp together into Frontier Park, built on the ruins of the old prairie.

Deep below our feet, seeds from a grassland that once stretched beyond imagination slumbered in the alluvium.

ACKNOWLEDGMENTS

I'd like to thank everybody who helped me with this book, offered advice, read early drafts, set me straight on science, loaned me gear, invited me to give talks, and otherwise supported my work, especially Jay Bredwell, Christina, Chloe, Sara and Bill Frazier, Max McCoy, Kim Horner McCoy, Shirley Braunlich, Ted Dace, Dawn Buehler, the Kansas River Guides—David Sain, Marcia Endecott, Gregory Zolnerowich, Bill Hughes, Lisa Grossman, Steve Garrett, Andy Schone, Jane Liebert, and everybody else plus their scary baby head—Karen Merikangas Darling and the University of Chicago Press for taking the book on, Patrick Dobson, Walter Schroeder, Ned Kehde, Barbara Higgins-Dover, Martin Lockley, Mike Murphy, Jim Herrell, Mack Louden, Bob Bramblett, Kelly Kindscher, Craig Freeman, Lisa Ball, Gordon Warrick, Casey Cagle, Betty Jane Cattrell, Kim Hogeland, Derek Helms, Kelley Chrisman Jacques, Joyce Harrison, Ellen Wohl, Tina Casagrand, Kristin Soper, Rex Buchanan, Jane Cotton, Jeff Jensen, Marcia Lawrence, Jake Vail, Kelly Erby, Sarah Bagby, Clenece Hills, Helen Hokanson, Bunny Smith, Teresa Fernandez, Bruce McMillan, Bryan Williams, Christopher Janssen, Steve Schnarr, Scott Mansker, Ron Klataske, Roger Boyd, Dan Flores, Joyce Wolf, Karen Matheis, Jennifer, Austin, and April Haight, Lawrence Public Library, the Raven Bookstore, Watermark Books, and others that I've missed.

And to my MR340 sisters and brothers: Just keep paddling!

GRASSLANDS AND RIVERS WITHOUT END

Information about organizations and places mentioned in the book as well as a selection of prairie preserves, river resources, and groups that work to protect grasslands and rivers. Organized by chapter.

CHAPTER 1—RIVERINE DREAMS

Chance Cemetery

Address: 189th and Golden Road, Linwood, Kansas
Cemetery and Kansas River overlook at the mouth of Stranger Creek.

CHAPTER 2—MISSOURI

American Prairie Reserve

Website: www.americanprairie.org
Address: 302 W. Main Street, Lewistown, MT 59457

Charles M. Russell National Wildlife Refuge

Website: https://www.fws.gov/refuge/charles-m-russell
Address: 333 Airport Road, Lewistown, MT 59457

First Peoples Buffalo Jump State Park

Website: https://fwp.mt.gov/stateparks/first-peoples-buffalo-jump
Address: 342 Ulm Vaughn Road, Great Falls, MT 59404

Giant Springs State Park

Website: https://fwp.mt.gov/stateparks/giant-springs
Address: 4803 Giant Springs Road, Great Falls, MT 59405

The Nature Conservancy in Montana

Website: https://www.nature.org/en-us/about-us/where-we-work
/united-states/montana

Upper Missouri Breaks National Monument

Website: https://www.blm.gov/visit/upper-missouri-river-breaks
-national-monument
Address: Between Fort Benton on US Highway 87 and Robinson
Bridge on US Highway 191

Upper Missouri National Wild and Scenic River

Website: https://www.blm.gov/visit/upper-missouri-national-wild
-and-scenic-river
Address: 701 7th Street, Fort Benton, MT 59442

CHAPTER 3—GRAND

Adam-ondi-Ahman

Address: 22379 Koala Road, Jameson, MO 64647

Dunn Ranch Prairie

Website: https://www.nature.org/en-us/get-involved/how-to-help
/places-we-protect/dunn-ranch-prairie
Address: 16970 W. 150th Street, Hatfield, MO 64458

Elam Bend Conservation Area

Website: https://mdc.mo.gov
Address: 590 Road, McFall, MO 64657

Grand River Grasslands

Website: https://mdc.mo.gov/your-property/priority-geographies
/grand-river-grasslands
Location: Northwest Missouri and Southwestern Iowa

Missouri Prairie Foundation

Website: https://moprairie.org

The mission of the Missouri Prairie Foundation is to protect and restore prairie and other native grassland communities through land acquisition, management, education, and research.

Prairie State Park

Website: https://mostateparks.com/park/prairie-state-park
Address: 128 N.W. 150th Lane, Mindenmines, MO 64769
Prairie State Park is Missouri's largest remaining tallgrass prairie landscape.

CHAPTER 4—KAW

Akin Prairie

Website: https://www.klt.org/akin-tribute
Address: On N. 1150 Road west of E. 1900 Road/County Road 1057
16-acre upland prairie in Douglas County.

Friends of the Kaw

Website: https://kansasriver.org

Ivan Boyd Prairie Preserve

Website: https://www.douglascountyks.org/county-parks/ivan -boyd-prairie-preserve
Address: 2011 N. 200th Road, Wellsville, KS 66092
Small prairie with Santa Fe Trail ruts next to Blackjack Highway Park and Battle Site.

Kansas Land Trust

Website: https://www.klt.org
The Kansas Land Trust protects and preserves lands of ecological, agricultural, scenic, historic, or recreational significance in Kansas.

Konza Prairie Biological Station

Website: https://www.nature.org/en-us/get-involved/how-to-help /places-we-protect/konza-prairie
Address: 100 Konza Prairie Lane, Manhattan, KS 66502
Konza Prairie is an ecological research station and one of the largest prairie preserves in the Kansas Flint Hills.

Lawrence Prairie Park

Website: https://lawrenceks.org/lprd/ppnc
Address: 2730 Harper Street, Lawrence, KS 66046
100-acre nature preserve with virgin tallgrass prairie and nature center.

The Nature Conservancy in Kansas

Website: https://www.nature.org/en-us/about-us/where-we-work
/united-states/kansas

Rockefeller Prairie at the KU Field Station

Website: https://biosurvey.ku.edu/rockefeller-experimental-tract
Address: Wild Horse Road and First Street, Lawrence, KS 66044

Tallgrass Prairie National Preserve

Website: https://www.nps.gov/tapr
Address: 2480 Ks-177, Strong City, KS 66869
Established on November 12, 1996, the preserve is the largest tallgrass prairie in the National Park system.

Wakarusa Wetlands

Website: https://www.bakeru.edu/history-traditions/the
-wetlands
Address: 1365 N. 1250 Road, Lawrence, KS 66046

CHAPTER 5—PURGATOIRE

Comanche National Grassland

Website: https://www.fs.usda.gov/recarea/psicc/recarea/?recid=
12409
Address: 1420 E. 3rd Street, La Junta, CO 81050

La Junta Tarantula Festival

Website: https://chooselj.com/tarantula-festival
Address: Downtown La Junta
Date: First weekend in October

The Nature Conservancy in Colorado

Website: https://www.nature.org/en-us/about-us/where-we-work
/united-states/colorado

Withers Canyon Trailhead

Website: https://www.fs.usda.gov/recarea/psicc/recreation
/recarea/?recid=12444&actid=50
Address: CR-25, La Junta, CO 81050

CHAPTER 6—SOUTH PLATTE

Carson Nature Center

Website: www.ssprd.org
Address: 3000 W. Carson Drive, Littleton, CO 80120–2983

Confluence Park

Address: 2200 15th Street, Denver, CO 80202

The Greenway Foundation

Website: https://thegreenwayfoundation.org

Pawnee National Grassland

Website: https://www.fs.usda.gov/recarea/arp/recarea/?recid
=32170&actid=34

Rocky Mountain Arsenal National Wildlife Refuge

Website: https://www.fws.gov/refuge/rocky-mountain-arsenal
Address: 6550 Gateway Road, Commerce City, CO 80022

CHAPTER 7—NIOBRARA

Cowboy Trail

Website: https://outdoornebraska.gov/parks/explore-trails/hiking
-biking/cowboy-trail
Address: Valentine, Nebraska

Fort Niobrara National Wildlife Refuge

Website: https://www.fws.gov/refuge/fort-niobrara
Address: 39983 Refuge Road, Valentine, NE 69201

Nature Valley Preserve

Website: https://www.nature.org/en-us/get-involved/how-to-help
/places-we-protect/niobrara-valley-preserve
Address: 42269 Morel Road, Johnstown, NE 69214

The Nature Conservancy's preserve on twenty-five miles of the Niobrara protects sandhills prairies and riparian forest.

Nebraska National Forest

Website: https://www.fs.usda.gov/nebraska
Address: 40637 River Loop Road, Halsey, NE 69142

Nebraska Star Party

Website: https://www.nebraskastarparty.org

Niobrara Wild and Scenic River

Website: https://www.nps.gov/niob
Address: 214 W. Highway 20, Valentine, NE 69201

Smith Falls State Park

Website: https://outdoornebraska.gov/location/smith-falls
Address: 90165 Smith Falls Road, Valentine, NE 69201

Valentine National Wildlife refuge

Website: https://fws.gov/refuge/valentine
Address: 39679 Pony Lake Road, Valentine, NE 69201

CHAPTER 8—MISERY

Arrow Rock State Historic Site

Website: https://mostateparks.com/park/arrow-rock-state
-historic-site
Address: 39521 Visitor Center Drive, Arrow Rock, MO 65320

Big Muddy National Fish and Wildlife Refuge

Website: https://www.fws.gov/refuge/big-muddy

Fort Osage National Historic Landmark

Website: https://fortosagenhs.com
Address: 107 Osage Street, Sibley, MO 64088

Kaw Point

Website: https://kawpointpark.org
Address: 1 River City Drive, Kansas City, KS 66117

Lewis and Clark Boathouse and Museum

Website: https://www.nps.gov/places/lewis-clark-boat-house-and
-museum.htm
Address: 1050 S. Riverside Drive, Saint Charles, MO 63301

Midwest Paddle Racing

Website: https://midwestpaddleracing.com

Missouri American Indian Cultural Center

Website: https://mostateparks.com/location/55530/missouris
-american-indian-cultural-center
Address: 32146 N. Highway 122, Miami, MO 65344

Missouri River Relief

Website: https://riverrelief.org

MR340

Website: https://mr340.org

The Nature Conservancy in Missouri

Website: https://www.nature.org/en-us/about-us/where-we-work
/united-states/missouri

Tucker Prairie

Website: https://mdc.mo.gov/discover-nature/places/natural
-areas/tucker-prairie

Van Meter State Park

Website: https://mostateparks.com/park/annie-and-abel-van
-meter-state-park
Address: 32146 N. Highway 122, Miami, MO 65344

MR340 EQUIPMENT AND SUPPLY CHECKLIST

Black carbon fiber North Star Magic 16-foot solo canoe

Obscenely light 54-inch bent shaft carbon fiber paddle

60-inch backup straight-shaft wooden paddle

Stadium-style canoe seat with extra cushioning

Ultralight black daypack

Rain poncho

Toilet paper

Lighter

Matches treated with candle wax

Chemical fire starter

Emergency mylar blanket

First aid kit

Waterproof blister tape

Two small rolls duct tape (one black, one silver)

Six tablets of Tylenol

Tums

Chemical ice pack

Gorilla glue

One gallon water jug

Camelback drinking hose threaded through the top of the canoe
    seat and attached to the life vest with Velcro

Insulated igloo cooler bag

Three 0.8-liter bottles of sugar-free Powerade sports drink

18 oz. premixed bottle of Hammer Perpetuem endurance formula
(plus extra powder in small bags)

Three boxes of vanilla bean organic nutrition protein shakes

Six Primal Spirit hickory-smoked vegan jerky sticks

Three Gopal's "Power wraps" made from nori and vegan jerky

Two hard-boiled eggs

Three small water bottles frozen rock-solid in a cooler of dry ice

Small bottle of hand sanitizer

Large bottle of SPF 50 spray sunscreen

Stick of white zinc sun lotion for nose

Small bag of sanitary wipes

Old-model Garmin eTrex GPS set to show speed in MPH

Velcro patches on the bow and stern to attach night navigation
lights

Marine knife velcroed onto life jacket to cut free from entangle-
ments if required

Map of Missouri River

Emergency referee's whistle

Yellow chemical light attached with carabiner to marine knife and
whistle on life jacket

Khaki-colored long pants

Green long-sleeved SPF shirt made of lightweight synthetic fiber

Neoprene paddling gloves

Inexpensive camo-colored hat with neck cover

NOTES

CHAPTER ONE

1 Quotation from Steven Buback, phone interview with the author, April 18, 2017.

CHAPTER TWO

1 National Park Service, *National Rivers Inventory*, accessed May 4, 2024, https://www.nps.gov/subjects/rivers/nationwide-rivers-inventory.htm.

2 *Wild and Scenic Rivers Act*, P.L. 90–542, October 2, 1968, https://www.rivers.gov/documents/act/complete-act.pdf.

3 Meriwether Lewis, *The Journals of the Lewis and Clark Expedition*, May 31, 1805, University of Nebraska Press / University of Nebraska-Lincoln Libraries-Electronic Text Center, https://lewisandclarkjournals.unl.edu/item/lc.jrn.1805-05-31.

4 Patrick J. Reilly and Kevin Froleiks, *Great Frontier: A Poorly Researched Musical about Lewis and Clark*, accessed December 8, 2024, https://www.show-score.com/off-off-broadway-shows/great-frontier-a-poorly-researched-musical-about-lewis-and-clark.

5 *The North American Journals of Prince Maximilian of Wied*, ed. Stephen S. Witte and Marsha V. Gallagher, 3 vols., trans. William J. Orr, Paul Schach, and Dieter Karch (Norman: University Press of Oklahoma, 2008–2012).

6 Witte and Gallagher, *North American Journals*, 1:235.

7 Lewis, *Journals of the Lewis and Clark Expedition*, May 31, 1805.

8 Dan Flores, *American Serengeti* (Lawrence: University Press of Kansas, 2016).

9 The Nature Conservancy, *Ecoregional Planning in the Northern Great Plains*

Steppe, Northern Great Plains Steppe Ecoregional Planning Team, 1999, accessed December 8, 2024, https://docslib.org/doc/4446418/ecoregional-planning-in-the-northern-great-plains-steppe.

10 Deborah E. Popper and Frank J. Popper, "The Great Plains: From Dust to Dust," *Planning* 53, no. 12 (1987): 12–18. Also available in American Planning Association, *The Best of Planning* (Chicago: American Planning Association, 1989), 572–77.

11 Daniel Kinka, "Evaluating the Effectiveness of Livestock Guardian Dogs: Loss-Prevention, Behavior, Space-Use, and Human Dimensions" (PhD diss., University of Utah, 2019).

12 This and following quotations from Daniel Kinka, phone interview by the author, March 11, 2022.

13 This and following quotations from David Crasco, phone interview by the author, April 16, 2022.

14 88th Congress, Second Session, Wilderness Act of 1964, Public Law 88-577 (16 USC 1131-1136).

15 *Missouri Breaks Scenic Recreation River: Hearing before the Subcommittee on Parks and Recreation of the Committee on Interior and Insular Affairs*, 87 (United States Senate, 92nd Congress, first session [1971] on S. 1405, a bill to establish the Missouri Breaks Wild and Scenic River).

16 *Missouri Breaks Scenic Recreation River: Hearing*, 85.

17 Bureau of Land Management, "Livestock Grazing on Public Lands," accessed December 8, 2024, https://www.blm.gov/programs/natural-resources/rangelands-and-grazing/livestock-grazing.

18 Amanda Eggert, "BLM Approves American Prairie Reserve's Bison Grazing Proposal," *Montana Free Press*, July 29, 2022, accessed November 20, 2022, https://montanafreepress.org/2022/07/29/blm-approves-american-prairie-reserve-grazing-application.

19 United Property Owners of Montana, accessed December 8, 2024, https://upom.org.

20 Legislature of the State of Montana, An Act Revising the Definition of Wild Buffalo and Wild Bison, and Amending Sections 81-1-101, 87-2-101, and 87-6-101, MCA, accessed May 4, 2024, https://leg.mt.gov/bills/2013/sb0399/SB0305_x.pdf.

21 Montana Department of Justice, "AG Knudsen to Federal Appeals Board: Overturn BLM's Illegal Bison Grazing Decision," accessed May 4, 2024, https://dojmt.gov/ag-knudsen-to-federal-appeals-board-overturn-blms-illegal-bison-grazing-decision.

22 Carrie Stadheim, "Seeking the Truth: Montana AG, Stockgrowers Appeal BLM's Bison Grazing Decision," *Tri-State Livestock News*, September 9, 2022, https://www.tsln.com/news/seeking-the-truth-montana-ag-stockgrowers-appeal-blms-bison-grazing-decision.

23 Kylie Mohr, "What's Getting More Expensive? Everything but Grazing Fees," *High Country News*, February 9, 2022, accessed December 4, 2024, https://www.hcn.org/articles/south-public-lands-whats-getting-more-expensive-everything-but-grazing-fees.

24 Lewis, *Journals of the Lewis and Clark Expedition*, May 29, 1805, https://lewisandclarkjournals.unl.edu/item/lc.jrn.1805-05-29.

25 Ellen Wohl, *Disconnected Rivers* (New Haven, CT: Yale University Press, 2004), 6.

26 United States Geological Survey,. "Cottonwood in the Missouri Breaks National Monument," 2005, https://pubs.usgs.gov/fs/2005/3132/report.pdf.

CHAPTER THREE

1 Jetta Carleton, *The Moonflower Vine* (New York: Simon and Schuster, 1962).

2 John Mitchell, *A Map of the British and French Dominions in North America, with the Roads, Distances, Limits, and Extent of the Settlements* (1755; Library of Congress, Geography and Maps Division).

3 George Lott, "The World-Record 42-Minute Holt, Missouri, Rainstorm," *Monthly Weather Review* 82 (1954): 50–59.

4 This and following quotations from Walter Schroeder, phone interview by the author, April 17, 2017.

5 Walter Schroeder, "Pre-settlement Prairie of Missouri," *Missouri Department of Conservation, Natural History Series #2* (Jefferson City: Missouri Department of Conservation, 1982).

6 This and following quotations from Steven Buback, phone interview by the author, April 18, 2017.

7 This and following quotations from Keith Bennett, phone interview by the author, July 11, 2022.

8 This and following quotations from Christopher Janssen,phone interview by the author, February 19, 2024.

9 Russell Schoof, "Environmental Impact of Channel Modification," *Journal of the American Water Resources Association* 16, no. 4 (2007): 697–701.

10 This and following quotations from Alexander Baugh, phone interview by the author, December 21, 2022.

11 Fawn Brodie, *No Man Knows My History: The Life of Joseph Smith* (New York: Alfred A. Knopf, 1945), 211.

12 Alexander Baugh, *A Call to Arms: The 1838 Mormon Defense of Northern Missouri* (Provo, UT: Joseph Fielding Smith Institute for Latter-day Saint History and BYU Studies, 2000).

13 This and following quotations from Edwin Howard, phone interview by the author, July 20, 2022.

CHAPTER FOUR

1 Tom Burns, *60 Years on the Kaw River* (Lawrence, KS: T. Burns, 1994).

2 This and following quotations from Barbara Higgins-Dover, interview by the author, June 18, 2022, Lawrence, Kansas.

3 Kansas Historical Society, "Lewis Dyche," accessed July 10, 2024, https://www.kshs.org/kansapedia/lewis-dyche/17799.

4 This and following quotations from Ned Kehde, phone interview by the author, March 19, 2018.

5 Burns, *60 Years on the Kaw River*.

6 Lonnie Doyle, "The Return of Iⁿʻzhúje ʻWaxóbe to the Kaw Nation," *Southwind Bulletin*, September 5, 2023, https://www.kawnation.gov/the-return-of-i%E2%81%BFzhuje-waxobe-to-the-kaw-nation/.

7 Stephen Harding Hart and Archer Butler Hulbert, eds., *The Southwestern Journals of Zebulon Pike, 1806–1807* (Albuquerque: University of New Mexico Press, 2007).

8 Hal Schramm, "Flathead Catfish—Homebodies or Migratory?," *Mississippi Sportsman*, January 1, 2021.

9 Ned Kehde, "Remembering Tom Burns: The King of the Kaw," *In-Fisherman*, March 20, 2103.

10 Burns, *60 Years on the Kaw River*, 17.

11 Quotation from Ben Neeley, phone interview by the author, July 22, 2022.

12 James H. Locklear, *In the Country of the Kaw: A Personal Natural History of the American Plains* (Lawrence: University Press of Kansas, 2024), 43.

13 Burns, *60 Years on the Kaw River*, 23.

14 Burns, *60 Years on the Kaw River*, 65.

15 This and following quotations from Heidi Mehl, phone interview by the author, February 2, 2023.

16 This and following quotations from Jerry Younger, phone interview by the author, February 2, 2023.

17 Waterkeeper Alliance, accessed May 18, 2024, https://waterkeeper.org/.

18 This and following quotations from Dawn Buehler, phone interview by the author, February 1, 2023.

19 This and following quotations from Mark Dugan, phone interview by the author, January 25, 2023.

20 Kansas Department of Transportation, "Watchers of the Valley: An Ethnohistory of Native Americans and the Wakarusa River Valley," in *South Lawrence Trafficway Construction, Kansas Turnpike to K-10, Lawrence: Environmental Impact Statement* (Lawrence: Kansas Department of Transportation, 1993).

21 Soren C. Larsen and Jay T. Johnson, *Being Together in Place: Indigenous Coexistence in a More Than Human World* (Minneapolis: University of Minnesota Press, 2017), 75.

22 Lawrence Bird Alliance, accessed May 18, 2024, https://www.jayhawk
audubon.org/baker-wetlands.

23 This and following quotations from Irene Unger, phone interview by the
author, January 5, 2023.

24 Meriwether Lewis, *The Journals of the Lewis and Clark Expedition*, June 26,
1804, University of Nebraska Press / University of Nebraska-Lincoln Libraries-
Electronic Text Center, accessed December 7, 2024, https://lewisandclark
journals.unl.edu/item/lc.jrn.1804-06-24.

25 Jay Senter, "Monument in Downtown Park Dedicated at City's 75th Anniversary," *Lawrence Journal World*, September 19, 2004.

26 From the National Day of Mourning Monument, https://plimoth.org/yath
/unit-5/national-day-of-mourning-monument-1998.

27 Quotation from Pauline Eads Sharp, phone interview by the author, February 10, 2023.

CHAPTER FIVE

1 Colorado Whitewater Association, "River Access," accessed June 2, 2024,
https://coloradowhitewater.org/river-access.

2 Cory Helton, *The Right to Float: The Need for the Colorado Legislature to Clar-
ify River Access Rights, 83* U. Colo. L. Rev. 845 (2012).

3 Harry E. Weakly, "A Tree-Ring Record of Precipitation in Western Nebraska," *Journal of Forestry* 41, no. 11 (November 1943): 816–19.

4 This and following quotations from David Augustine, phone interview by
the author, March 31, 2023.

5 Colorado State Forest Service, "Tackling Tamarisk on the Purgatoire River
Watershed: Restoring the Ecosystem Along the River," accessed May 24,
2024, https://csfs.colostate.edu/la-junta/lj-tamarisk-management.

6 Kevin R. Bestgen, Cameron T. Wilcox, Angela A. Hill, and Kurt D. Fausch,
"A Dynamic Flow Regime Supports an Intact Great Plains Stream Fish Assemblage," *Transactions of the American Fisheries Society* 146 (2017): 903–16;
compare with R. G. Bramblett and K. D. Fausch, "Fishes, Macroinvertebrates, and Aquatic Habitats of the Purgatoire River in Piñon Canyon, Colorado," *Southwestern Naturalist* 36 (1991): 281–94.

7 R. G. Bramblett and K. D. Fausch, "Variable Fish Communities and the Index of Biotic Integrity in a Western Great Plains River," *Transactions of the
American Fisheries Society* 120 (1991): 752–69. This and following quotations
from Bob Bramblett, phone interview by the author, March 31, 2023.

8 This and following quotations from Hank Guarisco, phone interview by
the author, March 31, 2023.

9 Justin O. Schmidt, "Instantaneous, Electrifying, Excruciating Pain: The Life
History of the Tarantula Hawk Is Similar to That of Many Other Solitary

Wasps. Its Sting, However, Is Not," *Undark*, May 18, 2016, accessed May 24, 2024, https://undark.org/2016/05/18/tarantula-hawk-wasp-sting-pain.

10 Karen L. Vigil, "Dinosaur Tracks Have Human History," *Pueblo Chieftain*, July 15, 1992, accessed December 8, 2024, https://www.chieftain.com/story /special/1992/07/15/dinosaur-tracks-have-human-history/8673204007.

11 This and following quotations from Martin Lockley, phone interview by the author, May 18, 2020.

12 Martin G. Lockley, Karen J. Houck, and Nancy K. Prince, "North America's Largest Dinosaur Trackway Site: Implications for Morrison Formation Paleoecology," *Geological Society of America Bulletin* 97, no. 10 (1986): 1163–76.

13 Martin G. Lockley, Richard T. McCrea, Lisa G. Buckley, Jong Deock Lim, Neffra A. Matthews, Brent H. Breithaupt, Karen J. Houck, Gerard D. Gierlinski, Dawid Surmik, Kyung Soo Kim, Lida Xing, Dal Yong Kong, Ken Cart, Jason Martin, and Glade Hadden, "Theropod Courtship: Large Scale Physical Evidence of Display Arenas and Avian-like Scrape Ceremony Behaviour by Cretaceous Dinosaurs," *Scientific Reports 6* (2016), art. no. 18952.

14 This and following quotations from Mack Louden, phone interview by the author, January 6, 2019.

15 This and following quotations from Jim Herrell, phone interview by the author, January 6, 2018.

16 Not 1 More ACRE! v. US Department of Army, Civil Action No. 08-CV-00828-RPM (D. Colo. Sept. 8, 2009).

17 Not 1 More ACRE! v. US Department of Army.

CHAPTER SIX

1 Environmental Protection Agency, "Climate Change Indicators: Snowpack," accessed June 1, 2024, https://www.epa.gov/climate-indicators /climate-change-indicators-snowpack.

2 Colorado.com, "Colorado Breweries: Defining the Craft," accessed April 2, 2024, https://www.colorado.com/articles/colorado-breweries-defining -craft.

3 CFS = cubic feet per second.

4 Francis Parkman, *The Oregon Trail: Sketches of Prairie and Rocky-Mountain Life* (New York: Macmillan, 1849), 67.

5 This and following quotations from Skot Latona, phone interview by the author, July 12, 2022.

6 This and following quotations from Ryan Aides, phone interview by the author, July 20, 2022.

7 P. A. Jones and T. Cech, *Colorado Water Law for Non-Lawyers* (Boulder: University Press of Colorado, 2009).

8 Open Water Foundation, "North Sterling Irrigation District," accessed

June 1, 2024, https://stories.openwaterfoundation.org/co/swsi-story-sp
-hydrology/visualizations/north-sterling-canal-storymap.html.

9   William E. Smythe, *The Conquest of Arid America* (New York: Macmillan
    Company, 1911), 161.

10 South Platte Regional Opportunities Water Group, "SPROWG Feasibility
    Study Report," accessed June 1, 2024, https://www.southplattebasin.com
    /documents/sprowg.

11 Rachel Carson, *Silent Spring* (Boston: Houghton Mifflin Company, 1962).

12 This and following quotations from David Lucas, phone interview by the
    author, March 14, 2022.

13 Edwin James, "Account of an Expedition from Pittsburgh to the Rocky
    Mountains, Performed in the Years 1819 and '20, by Order of the Hon. J. C.
    Calhoun, Sec'y of War: Under the Command of Major Stephen H. Long.
    From the Notes of Major Long, Mr. T. Say, and Other Gentlemen of the
    Party" (Philadelphia: Longman, Hurst, Pees, Orre, and Brown, 1823), 187.

14 James, "Account of an Expedition," 189.

15 Ellen Wohl, *Wide Rivers Crossed* (Boulder: University Press of Colorado,
    2013), 79.

16 This and following quotations from Rose Shirley, phone interview by the
    author, July 28, 2022.

17 Jeffrey A. Falke, Kurt D. Fausch, Robin Magelky, Angela Aldred, Deanna S.
    Durnford, Linda K. Riley, and Ramchand Oad, "The Role of Groundwater
    Pumping and Drought in Shaping Ecological Futures for Stream Fishes in
    a Dryland River Basin of the Western Great Plains, USA," *Ecohydrology* 4,
    no. 5 (September 1, 2011): 682–97.

18 This and following quotations from Dusty Barner, interview by the author,
    June 20, 2021, North Platte, Nebraska.

CHAPTER SEVEN

1   United States Census Bureau, *Nebraska: 2020 Census*, accessed December 31,
    2023, https://www.census.gov/library/stories/state-by-state/nebraska
    -population-change-between-census-decade.html.

2   Quotation from Amy Kucera, interview by the author, August 12, 2022,
    Smith Falls State Park, Valentine, Nebraska.

3   This and following quotations from Gordon Warrick, interview by the au-
    thor, August 12, 2022, Smith Falls State Park, Valentine, Nebraska.

4   National Park Service, "2022 National Park Visitor Spending Effects: Eco-
    nomic Contributions to Local Communities, States, and the Nation," Natural
    Resource Report NPS/NRSS/EQD/NRR—2023/2551, accessed December 7,
    2024, https://www.nps.gov/nature/customcf/NPS_Data_Visualization
    /docs/NPS_2022_Visitor_Spending_Effects.pdf.

5 Quotation from Jenna Bartja, phone interview by the author, June 14, 2023.

6 This and following quotations from Jessica Corman, phone interview by the author, April 26, 2024.

7 This and following quotations from David Manning, phone interview by the author, April 24, 2024.

8 Jennifer Dailey, *"Effects of Recreation in the Niobrara National Scenic River on Aquatic Macroinvertebrates"* (master's thesis, University of Nebraska-Omaha, 2023).

9 Quotation from Jennifer Dailey, phone interview by the author, April 16, 2024.

10 FNA Committee, *"Penstemon (Plantaginaceae),"* in *Flora of North America, vol. 17,* ed. *FNA Committee* (New York: Oxford University Press, 2019), *82–255.*

11 Quotation from Craig Freeman, phone interview by the author, July 5, 2024.

12 David Sutherland, "Historical Notes on Collections and Taxonomy of *Penstemon Haydenii* S. Wats. (Blowout Penstemon), Nebraska's Only Endemic Plant Species," *Transactions of the Nebraska Academy of Sciences* 16 (1988): 191–94.

13 This and following quotations from Mel Nenneman, phone interview by the author, May 18, 2023.

14 "Review of *Botany for High Schools and Colleges* by Charles E. Bessey," *American Naturalist* 14 (November 1880): 796–97.

15 This and following quotations from Greg Wright, phone interview by the author, July 12, 2023.

16 Henry David Thoreau, *Walden* (London: Macmillan Collector's Library, 2016).

17 88th Congress, Second Session. The Wilderness Act of 1964. Public Law 88-577 (16 USC 1131-1136).

18 Quotation from Matt Sprenger, interview by the author, May 26, 2023, Fort Niobrara National Wildlife Refuge, Valentine, NE.

CHAPTER EIGHT

1 Hiram Martin Chittenden, *History of Early Steamboat Navigation on the Missouri River: Life and Adventures of Joseph La Barge* (New York: Francis P. Harper, 1903), 288.

2 Leland R. Johnson, ed., "An Army Engineer on the Missouri in 1867," *Nebraska History* 53 (1972): 253–91.

3 John C. Luttig, *Journal of a Fur-Trading Expedition on the Upper Missouri: 1812–1813* (St. Louis: Missouri Historical Society, 1920), 33.

4 Meriwether Lewis, *The Journals of the Lewis and Clark Expedition*, June 22, 1804, University of Nebraska Press / University of Nebraska-Lincoln Libraries-Electronic Text Center.

5 *History of Saline County, Missouri: Carefully Written and Compiled from the Most Authentic Official and Private Sources: Including a History of Its Townships, Cities, Towns and Villages: Together with a Condensed History of Missouri, the State Constitution, a Military Record of Its Volunteers in Either Army of the Great Civil War . . . Biographical Sketches of Prominent Men and Citizens Identified with the Interests of the County* (St. Louis: Missouri Historical Company, 1881), 114.

6 Michael E. Dickey, *The People of the River's Mouth: In Search of the Missouria Indians* (Columbia: University of Missouri Press, 2011).

7 This and following quotations from Michael Dickey, interview by the author, June 25, 2023, Marshall, MO.

8 Reuben Gold Thwaites, *Jacques Marquette's Journal of 1673* (Cleveland: Burrows Bros., 1901).

9 Dickey, *People of the River's Mouth*, 4.

10 R. M. Morrissey, "The Terms of Encounter: Language and Contested Visions of French Colonization in the Illinois Country, 1673–1702," in *French and Indians in the Heart of North America, 1630–1815*, ed. R. Englebert and G. Teasdale, 43–75 (East Lansing: Michigan State University Press, 2013).

11 Dickey, *People of the River's Mouth*, 105.

12 Chris Luedke, MR340 paddler, accessed June 16, 2024, https://www.youtube.com/@340Paddler.

13 Fish and Wildlife Service, *Big Muddy National Fish and Wildlife Refuge Comprehensive Conservation Plan* (Bloomington, MN: US Fish and Wildlife Service Region 3, 2014).

14 This and following quotations from Tim Haller, phone interview by the author, July 6, 2023.

15 Fish and Wildlife Service, "Final Revised Recovery Plan for the Pallid Sturgeon," January 24, 2014 (Billings, MT: US Fish and Wildlife Service Region 5, 2014).

16 Curtis H. Freese, *Back from Collapse: American Prairie and the Restoration of the Great Plains* (Lincoln: University of Nebraska Press, 2023), 197.

17 This and following quotations from Wayne Nelson-Stastny, phone interview by the author, July 6, 2023.

18 This and following quotations from Bob Bramblett, phone interview by the author, March 31, 2023.

19 C. Phillip Baumel and Jerry Van Der Kamp, "Past and Future Traffic on the Missouri River," Staff General Research Papers Archive 11809 (Iowa State University, Department of Economics, 2003), 11–12.

20 This and following quotations from Richard Grenville, phone interview by the author, July 6, 2023.

21 Missouri Department of Conservation, "Early Prairies of St. Louis," *Missouri Prairie Journal* 3 (April 1981): 8–13.

22 Aldo Leopold, *A Sand County Almanac: And Sketches Here and There* (New York: Oxford University Press, 2020).

23 Max Oelschlaeger, "Ecological Restoration, Aldo Leopold, and Beauty: An Evolutionary Tale," in "Environmental Aesthetics and Ecological Restoration," special issue, *Environmental Philosophy* 4, no. 1 and 2 (2007): 149–62.

24 William R. Jordan III, "Some Reflections on Curtis Prairie and the Genesis of Ecological Restoration," *Ecological Management and Restoration*, July 21, 2010.

25 K. Kindscher and L. Tieszen, "Floristic and Soil Organic Matter Changes after Five and Thirty-Five Years of Native Tallgrass Prairie Restoration," *Restoration Ecology* 6 (1998): 181–96.

26 This and following quotations from Kelly Kindscher, phone interview by the author, July 6, 2023.

27 Chris Helzer, *The Ecology and Management of Prairies in the Central United States* (Iowa City: University of Iowa Press, 2009).

28 This and following quotations from Chris Helzer, phone interview by the author, July 3, 2023.

29 Quotation from Chris Newbold, phone interview by the author, July 3, 2023.

30 This and following quotations from Courtney Masterson, phone interview by the author, July 10, 2023.

SELECTED BIBLIOGRAPHY

This bibliography lists further readings on grasslands and rivers, as well as other material that I found helpful when compiling this work.

Aspen Journalism / Lindsay Fendt. "Major South Platte River Basin Project Would Maximize Reuse of Western Slope Water." The Water Desk, University of Colorado Boulder. May 5, 2020.

Atchison, Anne. "Place Names of Five West Central Counties of Missouri." Master's thesis, University of Missouri-Columbia, 1937.

Augustine, David, Ana Davidson, Kristin Kickinson, and Bill Van Pelt. "Thinking Like a Grassland: Challenges and Opportunities for Biodiversity Conservation in the Great Plains of North America." *Rangeland Ecology and Management* 78 (2021): 281–95.

Ballinger, Ian, and Ali Dover, dirs. *When Kings Reigned*. Film, 2017.

Baugh, Alexander. *A Call to Arms: The 1838 Mormon Defense of Northern Missouri*. Provo, UT: Joseph Fielding Smith Institute for Latter-Day Saint History and BYU Studies, 2000.

———. "The History and Doctrine of the Adam-ondi-Ahman Revelation (D&C 116)." Paper presented at Foundations of the Restoration: Fulfillment of the Covenant Purposes, 45th Annual Sidney B. Sperry Symposium, Brigham Young University, Provo, Utah, October 28, 2016.

Beaty, Kevin. "It's Not Looking Like a Year for Tubing the South Platte." *Denverite*, June 16, 2022. https://denverite.com/2022/06/16/its-not-looking-like -a-year-for-tubing-the-south-platte.

Bestgen, Kevin R., Cameron T. Wilcox, Angela A. Hill, and Kurt D. Fausch. "A Dynamic Flow Regime Supports an Intact Great Plains Stream Fish Assemblage." *Transactions of the American Fisheries Society* 146 (2017): 903–16.

Blumhardt, Miles. "Prairie Dog Plague Threatens Colorado Concert-Goers, Endangered Ferrets." *The Coloradan*, August 21, 2019. https://www.coloradoan

.com/story/news/2019/08/21/colorado-plague-outbreak-threatens-phish
-concert-goers-endangered-ferrets/2070256001.

Bogan, Michael A. *A Biological Survey of Fort Niobrara and Valentine National Wildlife Refuges*. Fort Collins, CO: US Department of the Interior, National Biological Service, Midcontinent Ecological Science Center, 1995.

Brady, Lawrence L., David A. Grisafe, James R. McCauley, Quinodoz H. Ohlmacher, and Kenneth Nelson. "The Kansas River Corridor—Its Geologic Setting, Land Use, Economic Geology, and Hydrology." Kansas Geological Survey. Open-File Report 98-2. January 1998.

Bramblett, R. G., and K. D. Fausch. "Fishes, Macroinvertebrates, and Aquatic Habitats of the Purgatoire River in Piñon Canyon, Colorado." *Southwestern Naturalist* 36 (1991): 281–94.

———. "Variable Fish Communities and the Index of Biotic Integrity in a Western Great Plains River." *Transactions of the American Fisheries Society* 120 (1991): 752–69.

Brink, Jack W. *Imagining Head-Smashed-In: Aboriginal Buffalo Hunting on the Northern Plains*. Athabasca, AB: Athabasca University Press, 2008.

Brodie, Fawn. *No Man Knows My History: The Life of Joseph Smith*. New York: Random House, 1971.

Bullinger, Tom. "Montana Refuge Divides Tribes and Ranchers." *High Country News*, April 28, 2017. https://www.hcn.org/issues/49-9/montana-prairie
-refuge-divides-natives-and-ranchers/.

Burns, Tom. *60 Years on the Kaw River*. Lawrence, KS: T. Burns, 1994.

Carson, Phil. *Across the Northern Frontier: Spanish Explorations in Colorado*. Boulder, CO: Johnson Books, 1998.

Coates, James. *In Mormon Circles: Gentiles, Jack Mormons, and Latter-Day Saints*. Reading, MA: Addison-Wesley, 1991.

Cumberland, Linda, and Robert L. Rankin, eds. *Kaánze Íe Wayáje: An Annotated Dictionary of Kaw (Kanza)*. Kaw City, OK: Kanza Language Project of the Kaw Nation, 2012.

David, Kenneth. *River on the Rampage: A Shocking Picture of Chaos in the Kaw River Basin, Which Presents a Constructive Way to Control a River*. New York: Doubleday, 1953.

Davies, P. M., C. J. Vörösmarty, D. Dudgeon, C. A. Sullivan, S. Glidden, C. Reidy Liermann, A. Prusevich, P. B. McIntyre, M. O. Gessner, S. E. Bunn, and P. Green. "Global Threats to Human Water Security and River Biodiversity." *Nature* 467, no. 7315 (2010): 555–61.

Dickey, Michael E. *The People of the River's Mouth: In Search of the Missouria Indians*. Columbia: University of Missouri Press. 2011.

Dobson, Patrick. *Canoeing the Great Plains: A Missouri River Summer*. Lincoln, NE: Bison Books, 2015.

Eggert, Amanda. "BLM Approves American Prairie Reserve's Bison Grazing

Proposal." *Montana Free Press*, July 29, 2022. https://montanafreepress.org /2022/07/29/blm-approves-american-prairie-reserve-grazing-application.

88th Congress, Second Session. The Wilderness Act of 1964. Public Law 88-577 (16 USC 1131-1136).

Falke, Jeffrey A., Kurt D. Fausch, Robin Magelky, Angela Aldred, Deanna S. Durnford, Linda K. Riley, and Ramchand Oad. "The Role of Groundwater Pumping and Drought in Shaping Ecological Futures for Stream Fishes in a Dryland River Basin of the Western Great Plains, USA." *Ecohydrology* 4, no. 5 (2011): 682–97.

Farmer, Jared. *Trees in Paradise: A California History*. New York: W. W. Norton, 2013.

Fendt, Lindsay. "A Closer Look at the High-Priced Plan to Store Water along the South Platte River." *The Tribune* (Greeley, CO), November 24, 2018.

Fenston, Jacob. "Mormons Returning to Northwest Missouri, 174 Years After 'Extermination Order.'" *KBIA Radio Website*, January 21, 2012. http:// kbia.org/post/mormons-returning-northwest-missouri-174-years-after -extermination-order.

Flores, Dan. *The Natural West: Environmental History in the Great Plains and Rocky Mountains*. Norman: University of Oklahoma Press, 2003.

Frazier, George. *The Last Wild Places of Kansas: Journeys into Hidden Landscapes*. Lawrence: University Press of Kansas, 2016.

Freese, Curtis H. *Back from the Collapse: American Prairie and the Restoration of Great Plains Wildlife*. Lincoln: University of Nebraska Press, 2023.

Friedman, Paul. *Valley of Lost Souls: A History of the Piñon Canyon Region of Southeastern Colorado*. Denver: Colorado Historical Society, 1988.

Grinnell, George Bird. "Cheyenne Stream Names." *American Anthropologist* 8, no. 1 (1906): 15–22. www.jstor.org/stable/659160.

Heat-Moon, William Least. *River-Horse: A Voyage Across America*. Boston: Houghton Mifflin, 1999.

Helzer, Chris. *The Ecology and Management of Prairies in the Central United States*. Iowa City: University of Iowa Press, 2009.

Hoogeveen, Nate. *Paddling Iowa: 128 Outstanding Journeys by Canoe and Kayak*. Des Moines: Otter Run Media, 2012.

Jenkinson, Clay S. "Seeing Through the Eyes of Maximilian and Bodmer." *Great Plains Quarterly* 30, no. 2 (Spring 2010): 135–37.

Johnsgard, Paul. *The Nature of Nebraska: Ecology and Biodiversity*. Lincoln: University of Nebraska Press, 2001.

———. *The Niobrara: A River Running Through Time*. Lincoln, NE: Bison Books, 2007.

Kansas Department of Transportation. "Watchers of the Valley: An Ethnohistory of Native Americans and the Wakarusa River Valley." In *South Lawrence Trafficway Construction, Kansas Turnpike to K-10, Lawrence: Environ-*

mental Impact Statement. Lawrence: Kansas Department of Transportation, 1993.

Kanza Language and Landscape. "Tour of Kansas Rivers and Places with Kanza Names." Accessed November 19, 2023. https://sites.google.com/site/kanza languageandlandscape.

Kehde, Ned. "Remembering Tom Burns: The King of the Kaw." *In-Fisherman*, March 20, 2013, 106–9.

Kendall, Justin. "The Search for the Garden of Eden." *Pitch Weekly Magazine*, September 6, 2007.

Kenner, Charles. *History of New Mexican–Plains Indian Relations*. Norman: University of Oklahoma Press, 1969.

Koziol, Liz, and James D. Bever. "The Missing Link in Grassland Restoration: Arbuscular Mycorrhizal Fungi Inoculation Increases Plant Diversity and Accelerates Succession." *Journal of Applied Ecology* 54 (2017): 1301–9.

Larsen, Soren C., and Jay T. Johnson. *Being Together in Place: Indigenous Coexistence in a More Than Human World*. Minneapolis: University of Minnesota Press, 2017.

Lockley, Martin G., Barbara J. Fillmore, and Lori Marquardt. *Dinosaur Lake: The Story of the Purgatoire Valley Tracksite Area*. Special Publication 40. Denver: Colorado Geological Survey, 1997.

Lockley, Martin G., Karen J. Houck, and Nancy K. Prince. "North America's Largest Dinosaur Trackway Site: Implications for Morrison Formation Paleoecology." *Geological Society of America Bulletin* 97, no. 10 (1986): 1163–76.

Long, Ben. "Can the Great Bear Survive Today's Great Plains?" August 18, 2020. https://www.themeateater.com/conservation/endangered-species/can-the -great-bear-survive-todays-great-plains.

Lucas, David. "Comprehensive Conservation Plan: Rocky Mountain Arsenal National Wildlife Refuge." Prepared by Rocky Mountain Arsenal Wildlife Refuge and US Fish and Wildlife Service, Lakewood, CO, December 2016.

Luning, Ernest. "River Towns: Denver; The South Platte's Dirty Past Promises a Pristine Future." *Colorado Politics*, September 18, 2021. https://www .coloradopolitics.com/denver/river-towns-denver-the-south-plattes-dirty -past-promises-a-pristine-future/article_8eced27a-1278-11ec-80b9-c7ec7 1939fae.html.

Manning, Richard. *Grassland: The History, Biology, Politics, and Promise of the American Prairie*. New York: Viking, 1995.

Matsch, Richard P. "Memorandum Opinion and Order in Civil Action No. 08-CV-00828-RPM." United States District Court for the District of Colorado, Denver, 2008.

Mattson, David. "Extirpations of Grizzly Bears in the Contiguous United States, 1850–2000." *Conservation Biology* 16, no. 4 (2002): 1123–36.

McCoy, Max. *Elevations: A River's Journey Through the Heart of America*. Lawrence: University Press of Kansas, 2017.

——. "Ordained by Hate." *Joplin Globe*, January 28, 2001.

McGonigle, Josephine Shirar. *Mankind Yields: The History of Franklin, a Thriving Town in Kansas Territory and the Role It Played in the Making of Our Country*. Lawrence, KS: Josephine McGonigle, 1978.

McKown, J. D. "Examination of the Kansas River from Its Mouth to Junction City, Kansas." Report to the Army Corps of Engineers, St. Louis, MO, January 8, 1879.

McLain, Robert, and Tad Britt. "Peopling the Picketwire: A History of the Piñon Canyon Maneuver Site." Champaign, IL: Engineer Research and Development Center, 2007.

McMillan, R. Bruce. "Bison in Missouri Archeology." *Missouri Archaeologist* 73 (December 2012): 77–134.

Miller, David. *The Complete Paddler: A Guidebook for Paddling the Missouri River from the Headwaters to St. Louis, Missouri*. Helena, MT: Farcountry Press, 2005.

Mlot, Christine. "Restoring the Prairie." *Bioscience* 40, no. 11 (1990): 804–9.

Moul, Francis. *The National Grasslands: A Guide to America's Treasures*. Photography by Georg Joutras. Lincoln: University of Nebraska Press, 2006.

Murphy, Dave. *Paddling Kansas*. Madison, WI: Trails Books Guides, 2008.

Nalewicki, Jennifer. "Inside a Remarkable Repository That Supplies Eagle Parts to Native Americans and Science." *Smithsonian Magazine*, September 2016.

National Park Service. "Economic Benefits to Local Communities from National Park Service Visitor Spending—2020." Report no. NPS/NRSS/EQD/NPSSDS-2021/1416. Washington, DC: US Department of the Interior, National Park Service, 2021.

Newbold, C., B. Knapp, and L. Pile. "Are We Close Enough? Comparing Prairie Reconstruction Chronosequences to Remnants Following Two Site Preparation Methods in Missouri, U.S.A." *Restoration Ecology: The Journal of the Society for Ecological Restoration* 28 (November 6, 2019): 358–68.

Nimz, Dale. "The Kiro Controversy and Flood Control in the Great Depression." *Kansas History: A Journal of the Central Plains* 26 (Spring 2003): 14–31.

Parks, Ron. *The Darkest Period: The Kanza Indians and Their Last Homeland, 1846–1873*. Norman: University of Oklahoma Press. 2014.

Perkin, Joshuah S., Keith B. Gido, Jeffrey A. Falke, Kurt D. Fausch, Harry Crockett, Eric R. Johnson, and John Sanderson. "Groundwater Declines Are Linked to Changes in Great Plains Stream Fish Assemblages." *Proceedings of the National Academy of Sciences of the United States of America* 114, no. 28 (2017): 7373–78.

Pierce, Don. *Exploring the Missouri River Country*. Jefferson City: Missouri Department of Natural Resources Division of Parks and Recreation, 1983.

Pitchford, Greg, and Harold Kerns. "Grand River Watershed Inventory and

Assessment." Report of Northwest Missouri Regional Fisheries Staff. St. Joseph, MO: Northwest Missouri Regional Fisheries, 2010.

Popper, Deborah E., and Frank J. Popper. "The Great Plains: From Dust to Dust." *Planning* 53, no. 12 (1987): 12–18. Also available in American Planning Association, *The Best of Planning*, 572–77. Chicago: American Planning Association, 1989.

Prendergast, Alan. "The War Next Door." *Westword* (Denver), February 22, 2011. https://www.westword.com/news/the-war-next-door-5111813.

Propst, D. L., and C. A. Carlson. "The Distribution and Status of Warmwater Fishes in the Platte River Drainage of Colorado." *Southwestern Naturalist* 31, no. 2 (1986): 149–67.

Purgatoire Watershed Partnership. "Purgatoire River Watershed Plan." Trinidad, Colorado. 2014.

Ruff, Candy L. "James Chance Recalled by Grandson during Chance Cemetery Visit." *Leavenworth Times*, September 18, 1988.

Saint Louis National Historical Company. *History of Caldwell and Livingston Counties*. St. Louis: Saint Louis National Historical Company, 1886.

Schmeisser, Rebecca, David Loope, and Joseph Mason. "Modern and Late Holocene Wind Regimes over the Great Plains (Central U.S.A.)." *Quaternary Science Reviews* 29, nos. 3–4 (2010): 554–66.

Schneiders, Robert. *Big Sky Rivers: The Yellowstone and Upper Missouri*. Lawrence: University Press of Kansas, 2003.

———. *Unruly River: Two Centuries of Change Along the Missouri*. Lawrence: University Press of Kansas, 1999.

Self, Matthew. "Biologists Battle Invasive Species Infesting Lower Kansas River." February 28, 2023. https://www.ksnt.com/kansasoutdoors/biologists-battle-invasive-species-infesting-lower-kansas-river.

Senter, Jay. "Monument in Downtown Park Dedicated at City's 75th Anniversary." *Lawrence Journal World*, September 19, 2004.

Shaw, Ethan. "The Grizzly Reclaims the Great Plains." July 2017. https://sciencing.com/the-grizzly-reclaims-the-great-plains-13400310.html.

Shields, Mike. "Tom Burns Is a Kaw River Legend in His Own Time." *Lawrence Journal World*, September 6, 1998.

Skinner, Charles. *Myths and Legends of Our Own Land*. Scholar's Choice Edition, 2015.

Snyder, Gary. *The Practice of the Wild*. San Francisco: North Point Press, 1990.

Stafford, Margaret. "Uprooted Town Moves to Higher Ground After Midwest Deluge of '93." *Los Angeles Times*, July 12, 1998.

Stimson, Henry L. "Purgatoire (Picket Wire) River, Colo." 10880 H.doc.387. Washington, DC: US Department of War, January 24, 1944.

Stubbendieck, James, Theresa R. Flessner, Charles H. Butterfield, and Allen A. Steuter. "Establishment and Survival of the Endangered Blowout Penstemon." *Great Plains Research* 3, no. 1 (1993): 3–19.

Tack, Jason D. "Beyond Protected Areas: Private Lands and Public Policy Anchor Intact Pathways for Multi-Species Wildlife Migration." *Biological Conservation* 234 (2019): 18–27.

Thompson, Craig. *Along the Kaw: A Journey Down the Kansas River*. Lawrence, KS: Craig Thompson, 2012.

United States Fish and Wildlife Service. *Endangered and Threatened Wildlife and Plants: Establishment of a Nonessential Experimental Population of Topeka Shiner (Notropis topeka) in Northern Missouri*. 78 FR 42702. Document number 2013-17087, 42702–18. July 17, 2013.

United States Department of Agriculture, Forest Service. McKelvie Geographic Area Rangeland Allotment Management Plans on National Forest System Lands on the Samuel R. McKelvie National Forest, Bessey Ranger District in Nebraska. 73 FR 61017. Document number EB-24408, 61017–18. October 15, 2008.

Weinberger, Eliot. *The Ghosts of Birds*. New York: New Directions, 2016.

Wilder, Ron. "Efforts Made 50 Years Ago to Alleviate Problems with Grand River." *Chillicothe News*, July 25, 2011.

Witte, Stephen S., and Marsha V. Gallagher, eds. *The North American Journals of Prince Maximilian of Wied*. 3 vols. Translated by William J. Orr, Paul Schach, and Dieter Karch. Norman: University Press of Oklahoma, 2008–2012.

Wohl, Ellen. *Disconnected Rivers*. New Haven, CT: Yale University Press, 2004.

———. *Wide Rivers Crossed: The South Platte and the Illinois of the American Prairie*. Boulder: University Press of Colorado, 2013.

Wood, Greg. "Paddling 340 Miles on the Missouri River." *Missouri Life Magazine*, November 2, 2017. https://missourilife.com/paddling-340-miles-on-the-missouri-river-2/.

Zhang, Sarah. "US Tallgrass Prairie's Microbial Past Revealed." *Nature* (2013). https://doi.org/10.1038/nature.2013.14069.

———. "What Happens to Meat When You Freeze It for 35,000 Years: A Gastronomic Investigation of Mammoth Feasts." *The Atlantic*, December 24, 2019. https://www.theatlantic.com/science/archive/2019/12/permafrozen-dinner/604069/.

INDEX